I0481765

Predicament

The scientific community has advanced many contributions and arguments in explaining the Universe, which eventually culminates in answering the question- Did the Universe architect itself or was it created?

ABOUT

About this inquiry, Did the Universe architect itself or was it created? It can be argued that if the Creator of this Universe unveiled to the world on a daily basis then mankind would have had a single consensus, which is not the current case. Hence resting on this premise there are clearly two sides to this argument: those in opposition to creation (Atheists and Naturalists) argue that their eyes or ears haven't witnessed any apparent communication from beyond. Those in favour claim not to require a daily virtual assembly with the Creator, but who rely on the obvious awareness in reference to His existence via insights deduce of the existing world. The opposing side has a united foundation while those in favour have varied stances and are broadly divided into as many parties as there are the number of world religions. Fully acknowledging the incompetence to represent all faiths, Masood Ahmed selects Islam to bear the brunt of modern assaults on faith & belief, filleting arguments sought by folly with the sword of truth- the Holy Quran. Agreeably Science is taken as the common denominator to base the final judgment. "ancient Astronomy—Allah's narrative" is an attempt to chronicle all pieces of narrative of the causes that bring our Universe into existence as detailed in the Quran. It is not merely ancient folklore, but a clear explanation of the processes invested in the architecture of the skies above—all from the Quran on what has been ratified by science and still less fathomed today.

SEQUENCE

The claims made in "ancient Astronomy—Allah's narrative" is one-of-the-kind with astronomical findings surfacing for the first time in this work. Charting from the pages of the Quran the book traces the Earth's beginnings and its primordial state, highlighting origins of its skies & atmosphere. It describes the formation of the solar system, the Sun & Moon bringing infallible scriptural premise to the table. Likewise, the origins, purposes, and descriptions of space objects such as meteors, asteroids, and comets are derived, furnishing eye-opening narratives with scripture. From beyond solar space, the book details the creation of stars, exoplanets, cosmic similarities comparable to findings on earth as told in the Quran. For the first-time scripture has been unearthed to talk about phases of stars, nebulae, red giants, white dwarfs, supernovae, pulsars and neutron stars.

The book brings out evidence from the Quran for galaxies, clusters, superclusters, filaments, and walls which is unprecedented in the field of theology. The discussion on the death of stars and the description of doom of galaxies as told in the Quran is spectacularly captured in this thrilling narrative. After laying a solid foundation from known sciences, the book then ventures to discuss a whole new world of interstellar communication and alien life—for which our known scientific understanding is yet to grasp but which has been narrated in the Quran.

FIRST ASTRONOMER

The history & evolution of astronomical sciences is unveiled. Father of the Prophets, the ingenious Abraham is proofed as the first successful astronomer in the history of human race. This revelation of Abraham as a peer Astronomer—who earned a covenant of long poured out blessings on humankind and to whom Astronomers today owe dues is at most justly liberating. Deductions of this fact are drawn independently from the Quran which to the author's knowledge has not been attempted before. Then feats in Egyptian astronomy are brought to light deducing contributions from those termed as God's Prophets. Following this, in the broader narrative of this book in glimpses, God's covenant with men of faith and what it means to human societies today is revealed.

ALIENS

The Jinn and the Angels have so far been seen in the lenses of scripture alone as nontrackable. This book declares bold statements which these beings have long been in due- to see them in the light of science, as creatures bound by the laws of physics and in an acclaimed research, it postulates their natures and capabilities. The thought sinks in of how small and feeble we are in comparison to the scales of this Alien reach. Scanning the promises of God to nations and His bestowal of hegemonies and legitimacy to rule. How are we performing amidst influential alien jinns and gatekeepers of heaven- the Angels? A mind-blowing telecast is drawn of jinns and the house of Angels in the backdrop of Cosmos we live surrounded in.

REVELATIONS

New astronomical findings from the Quran are consistently recorded as the pages are flipped by pleasantly surprising both the acceptors and the deniers. Some of them are: the exact sequence of formation of the lower atmosphere, Plasmasphere and magnetospheric skies which is not known in modern science, the fact that layers of atmosphere are structurally overlapped, pressure plummeting with altitude, refraction of light, moon's reflection of sunlight, comets and asteroids being made up of inflammable materials, water in celestial space, ionosphere rebounding, warfare via satellites, human ventures into outer space, sun as a non-special star among others – an illogical observation from 7th century view, sequence of creation of objects in solar system, speeds faster than speed of light which is unfathomable to have been thought of 14 centuries ago. The work calculates the Universe to be at least thirty many times greater in a vector than what we have fathomed. These and many more discoveries are unveiled in this work of 58 chapters grouped into 9 parts excluding introductory and concluding sections.

CLOSURE

Clearly, it is a book tailor-made to serve every Christian, Jewish and Muslim home. Arguing that our science-based inquiry still hasn't understood creator's plan on earth to test its inhabitants of freewill for belief and conduct, the discussion is engaged to understand anomalies in nature from weather to astronomical phenomenon befuddling modern minds, arguing that the 'God of gaps' argument was never an eastern saddle. Showing the Creator's powers of record keeping, "ancient Astronomy—Allah's narrative" leads you to the onset of doomsday. Traditional misunderstanding of the whole of universe experiencing a big crunch is argued as invalid and it brings out startling data from the Quranic purview to the contrary. It hypothesizes the true reach of doomsday and shows how it is entwined with the arguments of freewill and determinism. It describes the breakdown of cosmic space leading to the collapse of Sun and Moon, marking the end of planetary time and presenting the catastrophe surmounting earth and its skies in a hair-raising climax of the end as if the weather suddenly turned frigid.

VISIT

https://allahsnarrative.com/
Instagram: @masood.aussie
"Creation of Earth and Origins of life-Allah's narrative" is next.

ANCIENT ASTRONOMY—ALLAH'S NARRATIVE

MASOOD AHMED

Edited By

Muhammad Mansoor Sheik

Copyrights

Copyright © Masood Ahmed, 2017
All rights reserved.

United States Copyright Office Registration Number: TXu 2-078-572
Without the written permission of the Author, use of this material in any form or medium is illegal.
ISBN-13: 978-1983512537
ISBN-10: 1983512532

The Library of Congress has catalogued the paperback edition as follows:
Masood Ahmed, 1982—
ancient Astronomy—Allah's narrative / by Masood Ahmed
Library of Congress Control Number: 2018900357
CreateSpace Independent Publishing Platform, North Charleston, SC.

First release date, 15th January 2018, self-published.
Agency- CreateSpace an Amazon company, USA.

For permissions to use material from this product, publishing rights, franchise for sales of this book, for convertibles into other media and for translations please email- authorofancientastronomy@gmail.com with subject line- Permissions.

The scanning, uploading and distribution of this book via the internet or via any other means without the permission of the author/publisher is illegal and punishable by law. Please purchase only authorized electronic editions, and do not participate in or encourage electronic piracy of copyrighted materials. Your support of the author's rights is appreciated.

Acknowledgements

In acknowledgement of credits and contributions of various individuals and organisations. I owe immense thanks to the Editor of this work my cousin Muhammad Mansoor Sheik for his incalculable contributions towards a detailed manuscript appraisal; my indebtedness and thanks to GAIA space program for GAIA Sky Software. On similar token I present my thanks to NASA public domain information and media, ESA (European Space Agency), ESO (European Southern Observatory), NRAO (National Radio Astronomy Observatory), NOAA (The National Oceanic and Atmospheric Administration), WWF (World Wildlife Fund)-Canada, Wikimedia commons/Wikipedia authors: Kelvinsong; Brocken Inaglory; Kevin Jardine; Rursus; Rogelio Bernal Andreao; Andrew Z. Colvin; Mdf at English Wikipedia; atlasoftheuniverse.com, Arvind Paranjpye; Quran.com, corpus.Quran.com, Abdul Rahman for his comments on Corpus.Quran.com, Mazhar A. Nurani for hints to refer to classical Islamic works for Venus as the astronomical body alluded to in Quran.

My special thanks to innumerable teachers who helped imbibe me with enquiry and to Abdullah Yusuf Ali for his near immaculate translation of the holy Quran in traditional sense of his work. And to all those whom I have not mentioned here but who remain a key player in this project.

Illustration credits acknowledged here also appear on figure captions.

Table of Figures

Thanksgiving

In the honour of the *"dua"* (prayer) sought by my mother, Nusrath Khatoon and father, Mohammad Markiyar.

Preface

This book owes its making to the land of Red Kangaroos. Imagined, evolved, renditioned and then carved in Adelaide, SA—my heartfelt thanks to the Kaurna people, aborigines of Adelaide. After the many relentless discourses between belief and scepticism with my friend Peter Morris this book thus, as a result, was uniquely spun. As an immigrant to this land, I was blessed with few precious acquaintances with my first ever Atheistic faceoffs in Australia. Here, I learned from among agnostic Australians their understanding and take on—adverse effects of subscribing to any religion and belief in general. Eventually, which led to the inception of this book with Astronomy as the science that is explored to credit the existence of God the Creator.

During the beginning years of my university, an orientation program of the Quran organized by a few respected individuals at a nearby place of worship in my locality had shown me my journey of pondering over the Lord's book. I felt encouraged at home to attend those gatherings, for my uncle had presented to my father an English translation of the Quran by a great scholar and a British-Indian barrister, Abdullah Yusuf Ali. Other than my preferred rich in rhyme expression of Abdullah Yusuf Ali's translation, many other popular Quranic commentators were also studied in our gatherings. My parents often reminded me to attend these classes happening once a week as they saw me mostly dawdle away my time with friends over day-long cricket or similar pastime. Which was a clear impediment also on my university education. Their encouragement towards the book of God saw me steer clear from the usual growing in Bangalore where Bollywood and seasonal lures like of kite fights are still regarded as easy picks of amusement and play. As a change, I spent my time in that circle of eager to learn believers, who were serious about the teachings from the Lord's book, unbiased and educating themselves as they advanced, even if, a wisdom-catch occasionally came from a new and a young participant like myself. I immediately saw value being shown, not to me, but to the teachings that surprised us all from the Quran for they came out instructing us of our wrong proceedings which were long thought of as truthful elements in our societies. It also became clear that many generations of Muslims from the earliest times simply hadn't had the opportunity to reflect afresh on the book of God; for through many generations, societies were tailored to succumb to dictatorial roles of sparsely learned men.

For years struggling against my desires I efforted to ponder over the Quran, comparative religion, history of rise and fall of empires on portions of earth walked by Prophets of Abrahamic faiths and the challenges of representing science and religion together. My steady recital and research of the Quran opened doors to many truths and certainties. The verses related to various streams of sciences from the Quran were calling out to me for many determinations. I wanted to announce what I had discovered but I could not conclude a narrative like how I would on my fundamental science problems at work. After having worked for several years as an Engineer designing commercial airplane structures, I was empowered with a whole new discipline which helped me to ascertain truthful positions in theology. One lovely thing happened to me was I almost did not learn religion from the books of men. Beware! that you may learn wrong from me. The only book, I acknowledge to have read and attempted reflection other than books for my university exams was the holy Quran. Four years before now, around the time I prepared to apply for immigration to a world down under, I had already read enough material of the emergent writings on religion and modern science to assume a stage of my own intellectual fulfillment. Followed by a clear Quranic base, I was now geared to separate fact from fiction.

Then, a long friend of mine, upon his acknowledgment of the Quran's truth, I saw him wonder concerning artistry found in this grand universe. He sounded: Why is this universe so vast? I had no clear answer. On yet another occasion, this time in a serious debate with a sceptically positioned mind in Adelaide I was remarked, "look at the vastness of the Universe." Is it possible for all of this to have had a creator? Perhaps I hadn't considered such questions before. This became the turning point which made me embark on a journey to find answers. I felt humbled to crunch big answers that I hadn't yet structured enough in my knowledge to speak in clear discernible words. It was the beginning of bad time, upon my immigrating to a new home & life in Australia, whereby I had to clearly represent that which I believed in or simply chide away. If I wasn't incumbent to work on deciphering science of astronomy, the big narrative of this day & age; objectively from the Quran, despite it being found adept in making many inferences alluding to the truth of God's existence, I was to bear the brunt of an everlasting shame. I wished that this bad time would soon be turned into good time in this upright age of modern-day Astronomy.

My pondering over the Quran had made me discover that it had in it innumerable verses on science especially on the Cosmic front. But for that I was lacking a story line to narrate. I'm thankful that this was the task that my Lord motivated me to undertake. The computer came in a very handy tool. I express my indebtedness to two people, Muḥammad ibn Mūsā al-Khwārizmī- formerly Latinized as Algoritmi, a Persian mathematician, astronomer, geographer and scholar during the Abbasid Caliphate in the House of Wisdom in Baghdad and to Charles Babbage- the English mechanical engineer and a polymath considered as father of the computer, who invented the first mechanical computer in early 19th century. Further, under no reservations I express my deep indebtedness to the ambitious western-science contributions for their pioneering lead on the space-technology front, who have made my day bright, by God's grace. Thus, it convened that all verses on this subject from the Quran be separated for an exclusive reflection. On the earlier run, I had gotten skills of my Arabic grammar- a need kit. Now the time had arrived that I sincerely attempted to fight bias through the divine Quran. Allah, my Lord has thus helped me narrate His story on this most awaited subject, "ancient Astronomy—Allah's narrative" is my honour. I thank the Lord of the worlds much for this.

Introduction

If I were to polish my words, I will be praised by every insincere, ingrate; lest, I may suffer separatism. But my Lord's pleasure will suffice.

Gates of the sky were flung open for science discoveries, about the universe in which we dwell. No subject has caused more fascination and fury than the sheer material displays in the Cosmic artistry. As it is said our makeup is of star dust. It is probably true, that learning about astronomy can transform us in a deep way needed in modern times for us to become better at seeing the bigger picture from the narrowness of our busy lives. It unites us with every living creature seen on earth and many hidden aliens, along with aspects of amazement found in the wider Cosmos. Astronomy gives the awe of wonderment of Lord's artistry. Infusing in us the ability to concede the hand of the Lord in many comprehending and yet other mysterious astronomical sequences. Culling the popular memes such as religion and science are streams apart. And dismissing most ancient myths to usher knowledge to percept modern-day observed anomalies. Some will find this, as popularly said deeply frightening and others ineffably thrilling.

When we no longer look at names by their ethnic binding as rednecks[1] in states would do, as something from an alien world to smatter with; when we regard every concept clear in want of reception, as one which has had endured a long-tureenful history, needing just the lies of it be cleared, as a detective at a crime scene would do; when we would like to disbelieve in comportment of national cultures on societal framework of anathema predispositions, as associates of reason would do; when we are ever more sacrificial to help humanity for another pedestal leap to see light, as the summing-up of human want would, to promote truth; altogether, striking the similitude of the many forces in nature that ridge the beauty and glory, effacing unworthy traits; when we thus, view literature in intact-scriptures[2] from its needed simpler tale of telling natural wonders to ordinary men—does the proof of the Lord show, how correct it would be!

The author of the Quran- Our Lord, makes obvious statements that facilitated people of the seventh century to understand the Quranic expression of the natural world, and yet codified it for future generations to assume their respective reflections over it. The Quran was beyond doubt delivered by an All-Knowing, All-Wise, Lord—Allah. Allah is the Arabic equivalent of- 'The God' in English. Therefore, its claim that ambiguity cannot be found in it stands uncontested in the history of its manifestation. The Quran became first ever literary work in Arab culture and the only source of reference for Arabic grammar to this day.[3] The miraculous claim of the Quran by its language before the Arabs was unusual and that too when it was spelled from a man unlettered per their traditional records in Makkah, Arabia.[4] As it is a fact that Prophet Muhammad wasn't skilled, except for he excelled in mannerism among his tribesmen; neither, in the historical sciences, nor in sciences of the natural phenomenon of earth or the Cosmos. Thus, Quranic information elates as a miracle for people in present times

---

[1] Racists in USA are often called rednecks like bogans in Australia and New Zealand slang.

[2] The Quran is an intact-scripture. Original recitation is still intact in various Arabic scripts & calligraphies even though the Arabic script itself has evolved over centuries.

[3] Arabic-literature on britannica.com. An excerpt: "...the aftermath of its (the Quran's) revelation led to a lengthy scholarly process that traced its precedents and analysed the Arabic language system; as such, its revelation also needs to be viewed as the event that marks the initial stages in the recording and study of the Arabic literary tradition."

[4] Quran 29:48 And you (Muhammad) were not reciting before it any book, nor were you (able) to inscribe one with your right hand.

as it did previously. There is no Arabic equivalent script that the Quran can be made comparable to in resemblance and in information.

The nature of Quranic Arabic attains an unsurpassable elegance. As it stands, obviously, the Arabic poets utterly failed to emulate the challenge given to them to prove their claim of Muhammad self-producing it.[5] Arabic, then hadn't had a fully developed script, thus the Quran progressed as an oral recitation documented in memories of numerous men and women and later as the script evolved from its rudimentary past, it was rendered in newer scripts.[6] In fact, the Quran has been a recital committed to memory and sung in prayers ever since. The Quran has brought about exclusive information on the history of mankind and of its distant future, as among other signatures to examine its veracity from Allah for all times. And it has repeatedly recounted the end of those who suffered the Lord's displeasure. It also progresses to inform that mankind in haste has often disregarded Lord's high status and humane treatment toward other commons; for this reason—Almighty has brought about means in nature to show His furious face.

It would take expertise in related fields of sciences to give pertinent comments, such as of the learned geologists to comment, that if mountains were removed from the earth, it would result in a flattened earth surface, nil of slopes, troughs or valleys. It will that day be a different earth, talking about doomsday unfolding as discussed in chapter 9 in this book. In the Quran, there are more than thirty chapters named after the natural phenomenon and nearly a dozen of them named for Astronomical phenomenon or objects. Without ambiguity, it is telling something; clearly, it is claiming something. The Naturalists[7] assert that if there were a creator, He would have communicated or provided instructions introducing Himself to us as any rational being would do for pragmatic purposes. The absence of it, they conclude in haste, proves His nonexistence. "It is time to

---

[5] Quran 2:23 And if you are in doubt as to what We have revealed in stages to Our servant (Muhammad), then produce a Surah (chapter) like thereunto; and call your witnesses or helpers (If there are any) besides Allah, if your (doubts) are true.

[6] Those who commit the Quran by memory are called Hafiz (Pl. Huffaz), byhearts.

[7] In philosophy, Naturalism is the "idea or belief" that only laws and forces which are apparent operate in the world (as opposed to supernatural or spiritual). Its limits to accommodate the scope for unknowns or undiscovered forces is debatable.

learn from the Quran" as told by a white politician on an Eid celebration organised by Masjid Abu Bakar (Islamic centre of San Diego) in California, while in my long stint working for Goodrich Aerospace in Chula Vista in 2009. And numerous others perhaps hypocritically devouring thoughtless voters. With this question Peter Morris set me on the path of writing this book from my aimless wanderings in Adelaide. In Hotel Mantra we met after a holiday to India, I was barely seeing hope of achieving anything. Peter Morris sounded loud and clear to me: Masood, look at the Cosmos man; It is so huge! If there was a God who made that surely, He would have told us something of how He made it. I remarked He has. And Peter goes: Where is the evidence?

Overlapped-skies of planet earth has not come about in the Quran after modern science deduced conclusions of such facts.[8] Per critics, those were medieval sciences that which the Quran accumulated.[9] But substantiation of allegations is something that critics usually avoid. Every occasion the Quran is extolled for its miraculous nature in informing us about modern Scientific discoveries an equivalent rant is shovelled at it by its critics smearing medieval attribution of it. I would like to ask did the Greek philosophers as late as around the seventh century or perhaps earlier or any of the scholars of the renaissance had any idea about the sun's cruise in space?[10,11] It is only due to modern scientific enquiry launched in the twentieth century that it was made known in popular sense that the sun isn't static but travelling in space. This clearly shows Islamophobic propogandists such as WikiIslam is very much in the exact footsteps of former critics who ignored the Quran ignorantly. A seventh century document explaining modern discoveries—the Quran is an amazing text that stands up to the demands of people requiring proof to believe in the Lord of Cosmos. I dread

---

[8] Quran 67:3 (Allah) Who created seven Skies in overlapped layers: No want of proportion will you find in the creation of Most Gracious (Allah). Turn your vision again, do see you any flaw?

[9] In the history of Europe, the Middle Ages or Medieval Period lasted from the 5th to 15th century. It began with the fall of the Western Roman Empire and merged into the Renaissance and the Age of Discovery.

[10] Ancient Greek philosophy arose in the 6th century BC and continued throughout the Hellenistic period and the period in which Ancient Greece was part of the Roman Empire.

[11] The Renaissance a period in European history, from the 14th to the 17th century, regarded as the cultural bridge between the Middle Ages and modern history.

those who make requesting proof perhaps only an excuse to continue on roll down disbelief.

Wrong inferences drawn by subscribers of any faith which naturalists are quick to cash on, do not annul God's plausibility, the Creator, the narrator of His own saga. Proliferated Bible, also a scripture from Allah, has had this plaguing problem severely biting its acceptance in modern times. Subscribers too, if they realise on its worthy tokens that Christ for Lord's sake wasn't speaking American English or perhaps British, to remind Richard Dawkins of his misunderstandings that he is tied in. Thus, interpretations of men have undoubtedly found their way into the scripture which convincingly erodes its intactness, therefore lenience must be exercised with it. This is true for the science community too with matters of science. After all, we're the same men at day-break carrying work in various offices. In recent past, after months of speculation, scientists from the Advanced LIGO (Laser Interferometer Gravitational-Wave Observatory) project confirmed they had detected gravitational waves caused by two black holes merging about 1.3 billion years ago. Months of speculation is clearly rationalising! In those months of speculations in vigour, many enthusiasts sounded as though these waves were primitive gravitational waves of the Big Bang. That, which curiously had come timely at a time when in the dawn of science—men of light had built instrumentation for a timely catch with no further delay listening to the honourable Big waves, which instead turned out to be waves caused by merging of two black holes. [12] Such accidents numerous in every human endeavour happen even today despite the fingertip and mind-blowing stately technology. But spared Muhammad eons earlier in the seventh century to produce the noble Quran. Nay… they will not believe!

Presenting here in this work are many arguments against the naturalistic science narrative coming from the Atheistic foundations. Presented here are innumerable scriptural evidences in a cheese like convincing way to demonstrate that Almighty Lord has aforetime communicated his own chronicle of the vast and wondrous Cosmic artistry. Some of the very assertive, yet trivial arguments of

---

[12] Speculation and Prediction: Lawrence Krauss - Physics Made Easy published 13 Apr 2017 on YouTube video time: 51-55 minutes. And, Confirmed reports: Gravitational Waves: The Big Bang's Smoking Gun, By Miriam Kramer | July 11, 2016 on space.com.

naturalists are exposed as they failingly argue concerning the design of unimaginable to scale distances and sizes between stars and galaxies constructed in skies. The book notes that the earth has never stood abandoned by Allah as naturalists wilfully assert. Rather, it is served-to in immaculate ways for which evidence is laid expressly throughout this work. The unseen and undetectable community of Angels who work orders of God in the earth and in its skies, are a designated staff for this world to bring about matters concerning human affairs is a heavily discussed theme along with how Allah establishes communications with Angels encompassing our first Cosmic sky. The outcry of Allah's nonexistence is rebutted thoroughly on the premise of science of astronomy by refurbishing mind throbbing arguments from within the covers of the Quran. Thus, bottom lining on them that clearly—it is time for them to learn from the Quran!

For fourteen centuries the Quran contained secrets, many of which for the first time now are unlocked by this work. Frequently the book points to the impossibility of all this information from having been concocted from the mind of a human being – Muhammad. Then it puts forward challenges from the Creator Allah- to mankind on many aspects of Creation which makes us walk on the thinking path. It discusses that the cycle of rediscovery and discovery of the marvels of this Universe is all within the framework of the Creator and this cycle brings us towards the greater purpose of life – to recognize the Creator, which is the goal of this book.

Though this work is bespeaking to unveil full narrative on Astronomy as told in the Quran, its limits are my limits to knowledge. Inconsistencies in it are again my humble failures. And suppositions are my acumen to interpret the verses of my Lord, which I'm entitled to. It is an expression of my intellectual property bearing no truth greater to anybody's conscience. Having disclaimed, I begin narrating the beauty in the blanket surrounding planet earth. Its origins, and sequence of structural formation, its roles, then that which culminates its tale of breaking away from its purposes in chapter 9, then the solar system and its seventh century description which the Creator deemed fit to reveal, adaptive to current timeline of humanity for our ease of understanding spanning across times. Then the interface between earthly skies with the Cosmic expanse is discussed, elucidating similarities in Cosmos, with emphasis on God's patterns of creation as told in his revelations. The work is filled with dramatic revelations like on the evolution of astronomical sciences in world-history by identifying the first astronomers. The major seven Cosmic skies which has been a Semitic narrative is affluently detailed. Thus, discoursing the limits of human observation of the material existences in the Cosmos this work ushers many insights into the hidden sciences found in the Cosmos as made known in the Quran. This work also successfully drafts a couple of interesting deep-dives concerning alien lives. A case from the Lord of the worlds has been argued in my own words—powerfully enough with sufficient references from the Quran, an authentic source of truth supporting my narrative. Much other information dealing with subjects of known sciences are internet searchable for cited and further information.[13]

Further in this book I have brought forward the semantics of the scripture to elucidate Allah's narrative battling the hard to penetrate wall of bias bred for centuries surrounding it from its practice mates and critics alike. Therefore, making it easy for critics to refute authoritatively on a clear stage in the areas of their expertise. To achieve this the roots of Arabic language are explored wherever required. The breathtaking expanse of the Universe, its design and its mechanisms, the great and tiny details in its formation and the events and purposes leading to it all, are categorically described from this ancient scripture. This Quran, which though less ancient, having been revealed fourteen centuries ago, in the chronology of scriptures is the most recently revealed among major world faiths. The book unveils full narrative on Astronomical phenomenon of the observable universe and the patch of sky that has startled humanity since ancient times. Also, having thoroughly argued for ambiguity in star facts by way of analyses this book reasons science limits of astronomers as to how they are faring with assumptions. A great revelation is the history of Sirius and its role in shaping the architecture of the nearest sky. A whole chapter is dedicated to this and the discussion of high-density star cluster design embedded in the sky of this world. Evidence from the Quran is brought forward to reveal binary star systems such as Sirius system and its implosion—an information impossible for a 7th-century text to contain. The purposes for the grand scale of the universe as wide and as incomprehensible as it is and the discussion why it is at this Cosmic scale is herein thoroughly derived from the Quran. Having primarily covered the understanding of architecture imbedded in the Cosmos and the structure in its creation, by the

---

[13] Most Popular search destinations include: NASA Blogs and Wikipedia.

end of this book, one would realise how insignificantly we stand amidst this grandest of the grand creations. The view of religion would not be the same as it was before. This at least is my humble hope. A storyline is proposed for the many activities surrounding the workings of Angels with respect to worldly affairs from the jinn and the human reference frame. Make of Angels and Jinns is brought to the discussion. Then limits of their operations are discussed giving mind-boggling insights into the backdrop of human wonderment into the depths of the night sky.

On many occasions, the wrong narratives and erroneous renderings of verses of the Quran are pointed out and their corrected interpretations are put forward. The critics or clerics alike who have drafted their Quranic interpretations wishfully or rashly, in the process, have introduced numerous reading errors are the ones responsible for the allegation that the Quran propagates many unreasonable and unscientific views. To the critics the Quran reflects a pre-scientific, seventh century view of the natural world which is far from the truth as you shall see in this work. It's quite curious that those ardent on holding up such views are WikiIslam type modern critics who are nowhere near being any masters in the scholarship of Arabic or the Quranic language & its interpretation. Sincerely, just by being a Muslim or by advancing a claim to scholarship one does not automatically acquire abilities to interpret the Quran accurately whichever. Known in the communities as learned Muslims but within protected circles in my humble opinion are primarily responsible (on many occasions) for deceptive and credulous interpretations of the Quran. Perhaps our bitter fighting over reserving purist rights to its interpretation better explains the gravity of my assertion that we have in our past and present a very wrong idea of reflecting upon the Quran. Having said, this work facilitates a preamble of narrative on Astronomy from the Quran's perspective. Therefore, if at all any naturalist propagandist must show guts to the claim of truth it is easy to bring down a verse from a seventh century pen assisted by this narrative. The offer cannot be more lucrative. If I'm sincere to my claims of fighting bias before your evidence—when I won't buck, I'll at least be proved a counterfeit in my own conscience. Thus, readers are urged to discard previous mistaken notions of the Quran and Islam in exchange for truth. Unlike the claims of orientalist given to Islamophobia and its numerous critics, the Quran does not chide away from clearly explaining the scientific phenomenon for Metaphors or alternative meanings. As the Quran is the truth from Allah—it is vivid and disfavours phenomenology although we may suppose many different meanings of its verses, which is perhaps due to our lack of understanding of the many scientific phenomena. Wordings and content of Quranic verses often not only dismiss popular mythology and unscientific misconceptions of the time of its revelation but even the unscientific misconceptions of present times. Thence, perhaps to satiate naturalists, why would the Lord of the worlds deliver yet another parchment? Or even open their minds to faith? After He has seen that they have remained persistent in boast, ridiculing His Historical favour of communications, by not paying any attention to the scriptures that have come in succession.[14] Or even as they have omitted options from bearing in mind that they could be at fault and awhile a known intact-scripture the Quran is thus still guarded in its unadulterated format.[15] Open to be tackled, if they are in truth willing for a spell on winning any intelligent arguments.

---

[14] Quran 3:2-4 Allah - there is no deity except Him, the Ever-Living, the self-subsisting, Eternal. Revealed upon you (O Muhammad) the Book, in truth, confirming what came before it; as He sent down the Law (of Moses) and the Gospel (of Jesus) before this, as a guide to mankind. And He sent down the criterion (Quran, of wisdom between right and wrong).

[15] The holy Quran as a revealed recitation in classical Arabic (al-fusha). The Quran (recital) is unlike Gospel writers authoring books in the Bible. The Quran is chanted as taught by Prophet Muhammad verbatim given by Gabriel upon orders from Allah.

1 ARCHITECTURE IN EARTHLY SKIES

*Asking to **see** which could not be seen in seventh century was undoubtedly a science prophecy of what our Lord was considering providing us in technology.*

Quran 21:30 Do not the disbelievers see that the Skies and the Earth were a joined entity and We (Allah) separated them and made from water every living thing? Then will they not believe?

Nascent planet earth as it was being shaped with its skies still unformed and its time, is a detail worthy of praise. In an age gone by, the earth as it is now, and its atmosphere were a single earth mass, a primordial unit of creation before they were separated into a functional earth and its atmospheric skies. This detail captured in the Quran in the words, *"Do not the disbelievers see that the Skies and the Earth were a joined entity and We (Allah) separated them..."*, is an unprecedented verdict in the history of our science findings told in the Quran. It's worth is notable for its miraculous nature because of it becoming fathomable knowledge in our modern tell tales of scientific discoveries. The early details of formation of our planet are found explicitly versed in the Quran. It is not a surprise that a seventh century human effort to describe science were at fault. But it must surprise all critiques that a seventh century—alleged as a human-endeavoured scripture, is found narrating accurate scientific information, this primordial unit being the first among the many chronicled herein. The next inference in the verse *"...and We (Allah) separated them and made from water..."*

is more intricate that it synchronises findings of water soon after the formation of embryonic earth-mass when the separation of its skies had concurred. Further in the verse, another mesmerising revelation is made where it is telling of the creation of life, *"…and made from water life…"*, soon after the skies separation prompting first condensation of water on earth's surface, paving the way for origins of life.

> Quran 41:11 "Then" He directed towards the Sky while it was smoke and said to it and to the Earth, 'Come (into formation), willingly or by compulsion.' They said, 'We have come willingly'.
>
> 12. Thus, He completed them as seven Skies in two-days and engineered for each Sky its role…

The Quran in many places details the number of skies as seven. And here, it is found emphatic in its narrative; confirming that these skies were extracted by a separation from a single primeval earth-mass unit. The six-day creation story is Allah's narrative.[16,17] But the Quran does not teach a 24-hr based six-day time line, neither does any former scripture, it is perhaps only an errant interpretation by many fellows of old. The Arabic word <u>yaum</u> same as the Hebrew word, best describes—just a period. Which is used in the Quran in various contexts to depict, half a day, a day[18], thousand years, space-travel time of a day equal to thousand years[19] and fifty thousand years[20] and further an unnumbered period. In the unnumbered category is the day of judgement. In syntax, the same Arabic word (yaum) plural (iyyam) is used to depict many phenomena of space-time

---

[16] Quran 50:38 And We (Allah) did certainly create the blanket skies and earth and what is between them in six days, and there touched Us no weariness.

[17] Quran 11:7 And it is He (Allah) who created the skies and the earth in six days - and His Throne had been upon <u>water</u>…

[18] Plural of yaum is used to describe Ramadhan fasts during the day lasting between dawn and dusk talking about half part of a day. And in Quran 69:7 Which Allah imposed (punishments) upon them for seven nights and eight days in succession…; informing us of punishment on a people for seven days (iyyam) and eight nights with raging winds.

[19] Quran 32:5 He arranges (each) matter from the sky to the earth; then it will ascend to Him in a Day, the extent of which is a thousand years of those which you count.

[20] Quran 70:4 The Angels and the Spirit (Arch Angel Gabriel) ascend to Him (Allah's throne) in a Day which has a measure equal to fifty thousand years.

constructs in which the day of judgement is a unique page yet to unfold in real-time domain.

The duration of time in the context of creation of our planet earth, for obvious reasons isn't 24 hr based. Because, the planet itself was forming in the first early two days. It naturally ensues that each day alludes to a different time period. The Quran does infer a six-thousand-years period for the creation of planet earth and its skies when the six-day creation story of Quran 50:38 is read together with verse *Quran 22:47 "…verily a Day in the sight of your Lord is like a thousand years of your reckoning."* Not to be confused with the age of earth itself, what is being spoken of here is only the creation time frame. However even this is subject to change in the light of verse 32:4. The arguing is that after completing the creation process, Allah assumed His throne in the realms of His abode, *Quran 32:4 "It is Allah who created the skies and the earth and whatever is between them in six days; then He established Himself above the Throne…"* which had been over the water, opening a new dimension to the physical possibilities which could have facilitated this: *Quran 11:7 "And it is He (Allah) who created the skies and the earth in six days- and His Throne had been upon water…."* Since after completing the creation process of planet earth and its skies, His throne assumed a position in His abode different than the earlier parked position at creation site of earth, indicative of the past tense used in *"…His Throne had been upon water…."*

1. The creation time frame of six periods could be purely relative to the time frame of Allah's parked position at the creation site, we don't know. And the equation of a thousand years could be purely relative to His established position in His abode: *Quran 22:47 "…verily a Day in the sight of your Lord is like a thousand years of your reckoning."*
2. The other equally valuable view is Allah's assuming upon His throne clocks the equation of a day equal to a thousand years regardless of space and parked positions owing to concepts of general relativity. As the higher gravity factor slows down time at a site than at a site of lower gravity where time runs faster.

Understandably it is told in the Quran that the (kursi) pew of Allah extends over the skies and the earth which is placed upon His throne akin to a chair on a grand stage (see section 6 and 6.9 for full discussion), hence the relation for comparison of time is justified purely by the science of general relativity not knowing the

Brobdingnag scale and gravity of the throne itself. But as the understanding from the verses go, after creating the planet earth, its skies, and every other thing between them He assumed to His throne which had been over the water and then moved to His abode navigating away from the past tense in the cited verse.

So, the event of Allah's disembarking from His throne at the creation site is again suggestive of relativity in creation time, instituting His assumed position upon His throne as corresponding to the equation of a day equal to a thousand years. These physical scenarios in my opinion have left an open page for interpretation regarding the actual time frame invested for creation (or perhaps it could be six thousand years). Thus, arguably not disclosing the relation for disembarked time from the throne to our numbering of days during the creation process for comparisons, the creation time and that of the earth's destruction isn't explicitly numbered and is left to our efforts to estimate the truth of this matter by our science endeavours.[21] As can be noticed in the Quran, for other examples set by Allah to assess who among us best exerts our intellectual abilities to reason. The age of solar system or that of planet earth is not told in the Quran explicitly. Albeit, it does verdict clearly that man is the late comer to planet earth. But, the last among the creation of Allah, is yet to appear.[22,23] Man is clearly the late arrival to planet earth after an epoch of its historical existence had preceded. Awhile, when man wasn't even a thing of any talk, in the conference of Allah's assembly: Quran 76:1 *"Has there not been over Man a long period of Time, when he was nothing— (not even) mentioned?."*

In Quranic description of the creation of planet earth, a first two-day period was invested in the accretion of a nascent earth-mass.[24] Following this a four-day

---

[21] Quran 18:11-12 Then We draw (a veil) over their ears, for years, in the Cave, then We roused them, to test which of the two parties was best at calculating the term of years they had tarried.

[22] Quran 29:20 Say: "Travel through the earth and see how Allah did originate creation; so, will Allah produce a last creation: for Allah has power over all things.

[23] Quran 27:82 And when the word befalls them. We (Allah) will bring forth for them a creature from the earth able to speak to them, (saying) that people were, of Our (Allah's) signs, not certain (in faith).

[24] Quran 41:9 Say, "Do you indeed disbelieve in He who created the earth in two days and attribute to Him equals? That is the Lord of the worlds."

period was advanced to structure the topography.[25] And synchronous with the second stage of earth building a two-day period for the design of earth's blanket skies was arranged: *Quran 41:12 "Thus, He completed them as seven Skies in two-days and engineered for each Sky its role...."*

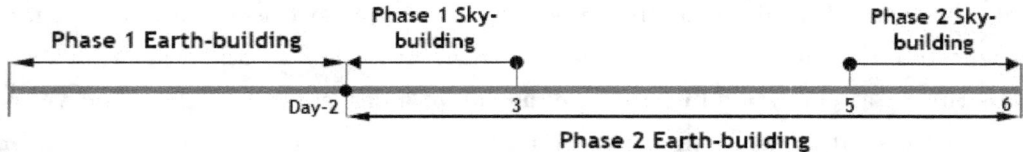

A diagram of the creation sequence of planet earth spread across six-days' timeline.

As part of second phase earth building, geological activity such as baking of earth's crust, cooling of its outer surface, breaking down of the crust into various plates, then displacement of these tectonic plates and formation of mountains had concurred. Flooded gases enveloping nascent earth-mass thus far had bellowed far and wide due to the calamitous early earth-building volcanic activity. It was a material-ready to become substance of the prime atmosphere. Heavily amassed by then, smoky constituents led to the formation of atmospheric sky *Quran 41:11 "Then He directed towards the Sky while it was smoke and said to it and to the Earth, 'Come (into formation)' ...",* where weather was organised as a sky layer as discussed in section 1.1. Awhile into the second phase, the tectonic plates were fissured to erect mountains and large displacements caused massifs at the colliding plate boundaries forming peaks & valleys across earth's crust which continued to the end of this phase. As the first weather in the beginning of second phase had resulted in waters which then condensed over the nascent earth's surface, microbial life at this stage concurred per Quran 21:30 (cited early in this chapter). The Quran advocates findings of ground based complex life-forms to emerge after the earth's surface had soaked enough fresh water causing rivers and streams to flow, dressing mountains, gorges, hills, and ranges with grasslands and plantations.[26] It is clear that the Quranic narrative

---

[25] Quran 41:10 And He placed in the earth firmly set mountains to a high altitude, and He blessed it and determined in it its sustenance (for creatures) in four days - in accordance with (the needs of) those who seek (Sustenance).

[26] Quran 13:3 And it is He (Allah) who spread out the earth and made firm mountains and (flowing) rivers: and fruit of every kind He made in pairs, two and two. He draws

rebukes Darwinian concept of evolution entirely based on natural selection as hedonistic. However, the Quran does not condone naturalists' presumptions of an instantaneous creation of species by the swing of a magic-stick to be the scriptural premise. But advocates a series-progression in the creation of species.

As the nascent planet had attained stability in its form of spherical shape, size and density (third day of creation time)—gravity at its full magnitude had resulted which was further perfected by the distribution of land masses as we know them today in the form of modern continents. This occurred on the final day of the total creation time- thus perfecting the design of all skies. At day three since geological processes were initiated for breaking of crust plates, the earth from within its heart was furiously geared up for a unique sky which was unleashed as a Magnetic sky. For this in deeper regions of earth's crust, the mantle and core were structured for an exacting performance. After launching in the third day this crucial sky was perfected on the final day as perfection of all skies concurred in total two-day periods: *Quran 41:12 "Thus, He completed them as seven Skies in two-days and engineered for each Sky its role...."* Crust plates as a result were moved and modelled to form new lands and the weather dressings of its topography were harmonized for providence to its many creatures on their course to future inhabit the earth. Bulk formation of mountains, flow of rivers and numerous allocations of resources were necessitated by the end of second phase of earth building. This has since then continued in the form of a sustained geological process, a decree of Allah for earthly beings to this day and to continue in the future.

In the second phase of planet building comprising of a total four-day period, formation of earth's skies was initiated, engineered and completed in part interims of two separate days. Sky building had synchronously begun with the earth's geological development marking the full realisation of gravity, of earth's atmospheric sky from its smoky constituents and the Magnetospheric rim being unleashed. The result of sums of water condensing, microbial life flourishing, earth's interiors coalescing, fissures gushing volcanic streams from deep within and oceans & seas sweltering had brought in the content and mechanisms for the

---

the night covering day. Behold, verily in these things are evidences for people of reflection.

completion of design of all Homospheric and Heterospheric skies under gravity and the magnetic lines of flux all from within the earth. When Allah commanded the smoky constituents and the earth-mass to coalesce, this resulted in the formation of seven-firmaments, as the Quran classifies them to be in an overlapped blanket surrounding the earth.

> Quran 67:3 (Allah) Who created seven skies in overlapped layers...

Atmosphere stacked by gravity had thus formed from smoke.

> Quran 41:11 Then He directed towards the Sky while it was smoke and said to it and to the Earth, "Come (into formation)" ...

Then as seen here in 67:3, more complex add-ons in the form of other skies emanated from within the earth, structuring to constitute two primary distributions of the atmosphere classified as: Homosphere and Heterosphere in a blanket enveloping it. Homosphere (see **Figure 1**) up to Karman line from earth's surface hosts homogenous atmosphere where weather takes place and in which Ozone as a unique sky layer is also imbedded. Beyond Homosphere, the Heterosphere at its very heart reigning with Plasmasphere, homes in Ionosphere and the Radiation belts. All the above being encased under the influence of gravity and under the overarching magnetic flux trapping the whole of atmosphere. Thus, the constituents and mechanisms which emanated from within the earth became its seven skies. This is the elementary description of Allah's artistry concerning the skies of planet earth. The following sections in this chapter will describe each of these skies in detail for you to wonder more at the Quranic narrative of each.

1.1 FIRST SKY

The homogenous part of atmosphere where there is a good mixture of gases is known as Homosphere. This in principle participates in refraction of light, facilitating visibility to earthly inhabitants. It is listed here as one out of the seven skies.[27]

Quran 79:27 Are you a more difficult creation or is the sky? Allah constructed it.

28. He raised its <u>sphere</u> and proportioned it.

29. And He darkened its night and extracted from it brightness (as day).

In Quran verse 79:28 shown here, the Arabic word سمك (samaka) meaning a pouter puffed up is implied for its spherical connotation. Pouter puffed up is a kind of pigeon able to inflate its crop considerably; from where the <u>spherical</u> inference is drawn. The overarching sky is not spherical per se, but increasingly a doughnut shaped region (a torus) with the near zero-hole (cusps) aligned with earth's magnetic axis (see **Figure 4**). The atmosphere isn't evenly spherically spread. In the use of attributing suffixes, such as "sphere" as in atmosphere, it is more to say in the figurative sense of a "sphere of influence." It is therefore interpreted as the spherical boundary of not just the atmosphere but the overarching spherical influence of planet earth's magnetic field. The Arabic word سَمْكها (samka-ha) is best used to describe the shape of earth's sky as a boundary region akin to a pouter puffed up. The pronoun ها (ha) is suffixed to it and connects the trilateral root سمك (samaka) of 79:28 to the sky talked in the preceding verse [79:27]. This is further given a description in 79:29 as *"darkened its night and extracted from it brightness (as day)"* this is a clear mention of bending of sun's incident light by way of refraction and reflection by atmospheric particles describing the process of visibility.

Imagine you were to walk on the surface of Moon stranded in space in darkness. Devoid of any atmosphere, even on the day side you can only see the surface and explicit objects in low visibility. In this hypothetical scenario if then a blanket of

---

[27] See **Figure 1** a key picture to understand important technical terminologies used in this chapter.

atmosphere is introduced, immediately it causes reflection and refraction of sun's light by gazillions of particles of the new atmosphere, extracting the light from the previous state of darkness, thus making it a bright visible realm. When the light from sun is not shown such as in the night side a state of darkness prevails. This occurs every day due to altercation of night and day. The choice of words used to describe this in the Quran is no mere coincidence, stating a miracle indeed.

A simple online translator can verify meanings of Arabic trilateral roots.

Troposphere which is distributed from earth's surface to an altitude of 20 kms within Homosphere has 75% of atmospheric mass and 99% of water vapor along with aerosols in a near spherical distribution. It extends nearly 20 kms in altitude in the tropics and 17 kms in mid altitudes and about 7 kms at the polar regions (not a perfect sphere but has a spherical inference). Likeness to the pouter of a pigeon when inflated is a befitting analogy. In this case troposphere accounts for 14.7 psi (pounds per square inch) atmospheric pressure with a total weight of 1.135×10^{19} lb (pounds).[28] The heavier molecules of atmosphere are pulled by gravity closer to earth's surface. Because of this the Homogenous atmosphere becomes concentrated near earth's surface and is thinned rapidly with altitude.

> Quran 6:125. Those whom Allah (in His plan) wills to guide; He opens their breast to Islam (submission to God); those whom He wills to leave straying; He makes their breast feel tight-constricted, as if they were ascending the sky: thus, does Allah place defilement upon those who refuse to believe.

In higher altitudes of atmosphere there are fewer molecules therefore the air pressure in upper horizons is lower. For example about 90 percent atmosphere is below an altitude of 16 kms from earth's surface; at about the summit of Mount Everest the air pressure is about 70 percent lower than it is at sea level. This means zealous climbers on top are breathing only 30 percent oxygen they would otherwise inhale at sea level. Regardless of oxygen needs, low pressure environments need pressure tight suits to help bodies fight constriction. It is one of the reasons why astronauts wear space suits. This is an amazing fact alluded in the Quran, the plummeting of pressure in higher altitudes of the sky is new

---

[28] Air pressure is in fact a measure of the weight of the molecules of atmosphere per unit area.

learning. To those who deny faith, Allah makes them feel their chest constricted to truth as if they were ascending the sky. This could be seen for the difficulty disbelievers express when their bias is challenged with truth. The Quran further informs that this sky hosts the weather phenomenon.

> Quran 2:164 ...and (Allah's) directing of the winds and the clouds controlled between the Sky and the Earth are signs for a people who use reason.

Most weather does occur in troposphere where there is most windy turbulence and where most of the cloud formation happens. But nonetheless, cloud formation has been documented through stratosphere up, well into altitudes of mesosphere wherein nacreous and noctilucent cloud formation phenomenon is observed respectively.[29]

> Quran 16:79 Don't they look at the birds hovered in the <u>atmosphere</u> of the Sky? None handles them except Allah. Indeed, in that are signs for a people who believe.

As read in verses of the Quran, it rightly points out that the clouds and birds hangout in the جو (jawu) meaning the <u>homogenous atmosphere</u> (Homosphere), suspended between the sky and the earth. The heterogenous atmosphere above karman line from an altitude of 100 kms and above cannot be classified as جو (jawu) due to rarefied strata of gases where clouds do not form, and neither can bird fly. Firstly, nailing the limits for the first sky to stretch up to an altitude which facilitates vision, hosts all earth-bound weather (winds and clouds), and hosts all known organic life including those of the Jinnkind discussed in chapter 7. This sky is alternatively known as the Homosphere or it can be named as the biospheric sky of planet earth where biosphere is the region where all ecosystems are found to zone life on earth's layer and in its sky. Bird species such as endangered Ruppell's griffon vulture and bar-headed goose have been reported to fly at lower levels of stratosphere classified by scientists as a second layer above troposphere. These birds reportedly overfly Mount Everest's summit (around 8.8

---

[29] Wikipedia: Polar stratospheric clouds or PSCs, also known as nacreous clouds, are clouds in the winter polar stratosphere at altitudes of 15,000–25,000 meters. And night clouds or noctilucent clouds are tenuous cloud-like phenomena in the upper atmosphere. They are made of ice crystals and are only visible in a deep twilight. Noctilucent roughly means night shining in Latin.

kms) conveniently. Also, bacterial life is found to survive in regions of stratosphere making it a part of biosphere.

The Quran's definition of its first sky is traced from earth's surface, in other words not necessarily from sea level; but, excluding oceansphere & lithosphere and while it traces the slopes of earth, lithosphere is being the rigid outer part of earth consisting of the earth's crust and upper mantle. This region along with oceansphere surrounding lithospheric earth's surface where life is found, just becomes a part of earth-based biosphere which can be understood per the following verse:

> Quran 14:24 ...example, of a good word is to a perfect tree, its root is firmly grounded, and its branches launched in the sky.

Secondly to conclude, the first sky per the Quran assumes from earth's surface tracing its mountainous, hilly slopes and extends up until what is identified as the Karman line at approximately the tropical altitude of 100 kms, where the aerodynamic lift flips into Kepler or centrifugal force. Because, of the amount of air contributing to aerodynamic lift plummeting significantly due to rarefaction of gases, the normal flight operations fail to come about. Also, it is a region of atmosphere where visibility significantly reduces due to rarefied atmospheric particles not refracting incident light from the sun as much as the particles well below do.

Scientists believe that a clear physical boundary where Homospheric-atmosphere جو (jawu) ends exists. Where the aerodynamics is replaced by astronautics. Thus, it is deduced per the Quran that the first sky is that which hosts all organic life, hosts all earth-bound weather (winds and clouds), stretches up to an altitude of Karman line, facilitates vision, thus making it the primary sky of planet earth. It hosts all life and the many essential natural phenomena directly influencing life. It encompasses the sub-classification of troposphere, stratosphere, mesosphere and lower thermosphere. Thus, for functionalities presented in the above narrative, the Quran terms this layer of the sky as a distinct sky layer called herein as the Homosphere. Also, per the Quran Homospheric sky houses another functional sky within it known as the Ozone, discussed in section 1.4.

1.2 SECOND SKY

The region of sky beyond Homosphere which is above the Karman line stratified due to rarefaction of gases is known as Heterosphere (see **Figure 1**). This region extends through thermosphere and exosphere seamlessly becoming free space at the edges of Magnetopause of Magnetosphere. This sky is herein termed as Heterosphere, reckoned as the second out of the seven skies.

The core of Heterosphere is Plasmasphere. Hence, though these are different names, Heterosphere is mainly called for Plasmasphere in this work. Essentially the whole of Heterosphere has plasma smeared into it, including plasma makeup of Ionosphere, another unique sky which the Plasmasphere encapsulates. Ionosphere shares a geometrical interface with Homosphere below tightly capsuling it. Plasmasphere, usually contains dense and cold plasma from above the peripheries of Ionosphere. Technically Plasmasphere, the oval shaped grey bubble region encompassing the planet earth is identified to extend protection by preventing high energy particles from sun crossing into near earth space (see **Figure 2**). And it resides in the inner regions of Magnetosphere. In addition to Plasmasphere, Heterosphere houses Ionosphere and Radiation belts which are discussed in sections 1.5 and 1.6 respectively. Plasma based region is an ongoing research area and nothing concrete is understood thus far on the exact functioning of plasma widely distributed and stratified in various massive proportions throughout Heterosphere. One observation is that, Plasmasphere is observed to source plasma from lower atmosphere; playing an important role for the build of Heterosphere. A sharp boundary around Magnetopause separates the plasma of Heterosphere from the sun dominated region known as heliosphere. This interface region is popularly known as the bow-shock shown as the rimmed donut shaped flux-lines of Magnetosphere **Figure 3** whose exteriors interact with sun's heliosphere on the dayside compressing the flux-lines significantly under the bow-shock pressure coming from the sun (see **Figure 4**).

There is no boundary at which Heterosphere abruptly ends but it is progressively thinned away at higher altitudes to eventually merge with free space. Here the chemical composition varies with altitude as distance between particles is large and they move greatly without colliding. This allows the gases to stratify by molecular weight with the heavier gases such as oxygen and nitrogen present only near lower Heterosphere. The upper Heterosphere is composed almost entirely of hydrogen being the lightest element.

Quran 52:5-6 And (by) the roof raised high. And by the sweltering Ocean.

The Arabic phrase in verse 52:5 here والسقف المرفوع (wa-saqhafi al-marfuaā) unconditionally brings out a new classification. Which means high in altitude constructed roof. The Quran in the above set of verses pin points an important relation. In that it alludes to the formation of Heterosphere: firstly, up to a very high altitude and secondly, due to its context implying to sweltering of oceans under rife volcanic activity. A critical component in the build of this sky layer which is implied to have directly resulted from water bodies sweltering due to dynamics of earth's volcanic crust. The Quran implies this subtle phenomenon to directly contribute to the elevation of plasma in the Heterosphere. Opinions of experts researching this area of sky are a worthy note.

Excerpt[30], from ongoing research: "…researchers are studying whistler waves triggered by **volcanic lightning** *to elucidate the structure of the earth's Plasmasphere. Which will eventually help scientists get better at anticipating the impact of space weather events. Lightning creates bursts of very-low-frequency radio waves – known as atmospherics, or "sferics" for short – and these electromagnetic waves sometimes penetrate the Ionosphere and make it into the Plasmasphere, where the sferics whizz between hemispheres through ducts of plasma where they become whistler waves. The passage of whistlers between the hemispheres provides a rare opportunity to investigate the Plasmasphere without the need to launch expensive space probes …"*

In the light of verses 52:5-6 which talks about the sky as a high in altitude constructed roof in context of sweltering oceans, the inference is thus drawn that Heterospheric sky per the Quran is that which includes Plasma, most of the thermosphere and all of exosphere. To conclude this whole distribution of primary atmospheric layers: Homosphere plus Heterosphere is said to block the deadly gamma radiations from entering earth's space thus giving the opportunity for life to safely begin.

---

[30] Title search: How volcanic lightning is helping to demystify the earth's Plasmasphere on phys.org.

1.3 THIRD SKY

The magnetic flux emanating from within the heart of the earth from its inner most regions of molten Iron core is the overarching field area which encompasses both Homospheric and Heterospheric part of atmosphere and is famously known as Magnetosphere (see **Figure 3**). Magnetosphere is functionally seen as a distinct sky overlapping all other skies but the satellite sky to its full depth on the day side of the sphere of influence. This is identified in the Quran as one of the seven skies, listed here at the third place.

As seen in the previous section, Plasmasphere along with overlapped radiation belts forms the inner most part of Magnetosphere (see **Figure 2**). The cold plasma of Plasmasphere trapped within earth's magnetic field lines is known to co-rotate with the earth. These field lines emanating from deep within the earth's Iron core are an astounding fact alluded to in the Quran as we shall see in Quran 41:11-12.

> Quran 41:11-12 "Then" He directed towards the Sky while it was <u>smoke</u> and said to it and to the Earth, "Come (into formation), willingly or by compulsion." They said, "We have come willingly."

> Thus, He completed them as seven Skies in two-days and engineered for each Sky <u>its role</u>. And (also) We (Allah) adorned the Sky of this world with lamps as protection. That is the determination of the Exalted in Might, the Knowledgeable.

The interpretation of Magnetosphere is derived from the above verses as it is found clearly elucidated. Words in verse 41:11 such as the atmospheric sky were once smoke clarify for the gaseous content surrounding early embryonic earth. Just prior to the event of its structural formation the volcanic smoky gases had engulfed early earth. Allah commanded this cover of smoky gas and to the earth mass to coalesce into a blanket design. The gaseous envelope's and earth's response to form together as spoken in the Quran, "willingly coming together" resulted in the atmosphere (Homosphere and Heterosphere) from what was primeval smoke to be enclosed within the blanket rim of Magnetosphere. As the earth's willingness to work together to enact Allah's command came about the Magnetic sky was unleashed by a powerful core operation from deep within earth's heart. If this Quran was not from the Lord of creation—Allah, it was far simple that this stark piece of information would be absent in the narrative alleged as human work. Presence of this info only confirms Allah's existence and communication of truth to mankind.

In action the night side of Magnetosphere extends its boundaries to about 1000 earth radii which is around six-million kms. Our satellite moon orbits the earth at about 60 earth radii, less than half a million kms. However, as seen from **Figure 3** and **Figure 4** the day side of it gets significantly distorted due to solar activity. The solar pressure compresses it to about 65,000 kms. The Magnetopause is the boundary between the planet's magnetic field and the solar wind. Magnetosphere encompasses Plasmasphere in its very interior regions and at its exteriors it powerfully bow-shocks the snares of the sun; this is the most revered of all—the Magnetospheric sky.

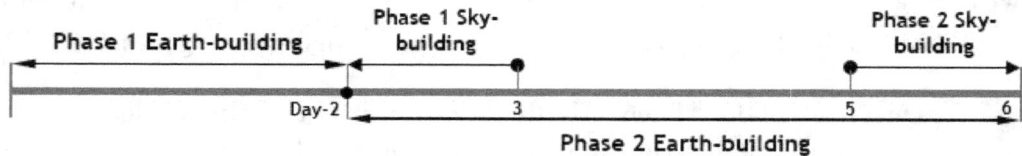

A diagram of the creation sequence of planet earth spread across six-days' timeline.

Most importantly, the Magnetic sky's formation time alluded to in the Quran is marked to commence synchronously with the earth's geological process and became perfected with plate tectonics, which commenced and concluded at the start and finish of second phase of earth building as indicated in *Quran 42:11* *"'Then' (after first phase of earth building) He directed towards the Sky while it was <u>smoke</u>...."* Per the Quran the sky building concurred in split phases, few skies resulted on the very first day and others along with the first ones were perfected in the last day of the second phase of earth building: *Quran 42:12 "Thus, He completed them as seven Skies in two-days...."*[31] Therefore, Magnetic sky's first formation was thus accomplished on the same day at the commencement of the second phase preceding the mountain formations as told: *Quran 42:11 "...and said to it and to the Earth, 'Come (into formation), willingly or by compulsion.' They said, 'We have come willingly.'"* Magnetic sky thus dropped its imprint as the magma in the rocks cooled, as is observable in the alignment of Iron-bearing minerals in the rocks recording its dynamic birth marks as changing field lines

---

[31] A concept well elucidated in this chapter in forthcoming sections.

and strength in the various mountains on earth. This is how the Quran surfaces to stage illuminating the truth!

Further it is read in the verse that roles to each sky were assigned: Quran 42:12 *"Thus, He completed them as seven Skies in two-days and engineered for each Sky its role."* It is telling that Allah engineered each sky to perform its functions. If the claim that much is being read into these verses of the Quran on the scientific front has any validity then such critics only need to substantiate this bit of the verse 41:12 which says: Allah, after finalising them as seven distinct layers, has assigned them their respective roles. If the Quran were not informing of this fact, that the earth has different layers of skies and roles then critics could have had a sense of winning composure. Alas! But critics now do shoulder the burden to explain why time and again the Quran has alluded to earth's skies being a plural number of seven than just a singular firmament. A single firmament indeed was all that was seen and observed in the middle ages up until modern scientific discoveries were made. There was no concept of skies being researched during medieval times but, rather all they saw was just perhaps only one sky.

1.4 FOURTH SKY

Located in lower regions of Homospheric atmosphere is a unique protective sky layer known as Ozone. Ozone is also functionally seen as a distinct sky overlapping through regions of troposphere and stratosphere. Identified in the Quran for its protective covering against harmful radiations as one out of seven skies, listed here in fourth position.

> Quran 2:22 (He) who made for you the Earth as couch and the Sky as canopy. And sent down from the Sky, rain and brought forth thereby fruits as provision for you...

The Arabic word بناء (binaa) translated as a canopy is descriptive of the basic build of an elemental structure from a seed crystal. Like a building erected from brick elemental units. Overlapped well within the Homospheric sky- this essential protective gear is built from elemental di-oxygen and tri-oxygen. The Quranic narrative normally carries pronouns to bring forward nouns within and across verses. As an exception in verse 2:22 cited here, the noun for sky is repeated a second time instead of succeeding it with a pronoun only to invoke an alternate function of a distinct sky layer. At first instance the noun i.e. sky, is implied as a canopy or a shelter. Then in the second occurrence it is alluded to bring down weather. Thus, this verse is indicative of describing functions of two different skies that are in same regional overlap. In our scientific understanding, Ozone structured between upper troposphere and lower stratosphere gives protection from the damaging UV radiation (shorter wavelengths) which could result in skin burns and DNA mutations, however UV waves of slightly larger wavelengths reach earth for many benefits as well (such as florescence based navigation for birds and insects). In a nutshell if not for Ozone, life would not be safe and flourishing on earth. Thus, Ozone is an essential layer for safe life. While the Homospheric sky and in specific the layer the troposphere brings down the weather phenomenon.

> Quran 40:64 It is Allah who has made for you the Earth a place of settlement (or life) and the Sky as cover (or canopy). And dressed you up in best of forms and provisioned you with good (organic) things. Such is Allah, your Lord; thus, blessed is Allah, Lord of the worlds.

Here in this cited verse, the Arabic word بناء (binaa) is again in analogous syntax as in verse 2:22, implying to shelter as in a cover or a canopy. In both citations, the same connotation is reflected in the syntax which is to build the case for a life nourishing and caring sky layer. For life to evolve securely, it needs a safe blanket or a protective sheath otherwise it is vulnerable to hostilities and would not flourish well. Both verses quoted in this context first imply the earth to be a couch and as a place of settlement for life and then progress to imply the sky as the canopy to nourish floral and faunal life. Both verses mentioning this sky layer imbed details of this very specific context which tell about its structural formation as having been concurred to support the emergence of higher life. As the verses then continue to elaborate on progression of sentient species such as trees bearing fruits and growth of its faunal eaters.

> Quran 67:3 (Allah) Who created seven Skies in <u>overlapped layers</u> ...

Here in verse 67:3 the Arabic word طبق (taubaqh) means <u>contours of dishes</u>. It is used as an adjective in the superlative form in the verse to describe structural overlapping of the seven skies. Thus, Ozone per the Quran descriptive of the Arabic word بناء (binaa) is the canopy built by structural brick elements such as dioxygen and trioxygen placed overlapped within Homosphere to protect life, it is thus acknowledged as the earth's natural sunscreen, absorbing and blocking most of the incoming harmful radiation from the sun in the UV range ensuring safe plant, animal, and Jinnkind (alien to us) life forms.

1.5 FIFTH SKY

The charged plasma present in the interface altitude of Homospheric sky overlapped within Heterospheric sky is the Ionosphere. It plays prominent role in atmospheric electricity and long-distance radio propagation due to ionization of atmospheric gases by the incoming sun's radiation energy. This layer of plasma extends across thermosphere and into the region of exosphere. Classified in the Quran by its unique possessive attribute to return radiation pressure and radio frequencies, it is listed here fifth in order as a unique sky.

> Quran 86:11-12 By the <u>Sky that rebounds</u>. And by the earth prone to fissures (or cracks or splits).

In the verses cited above an amazing relation is brought forward concerning the Ionospheric sky's ability to rebound radiation pressure and the earth's ability to quake. It is more amazing because this radiation pressure is released due to fissures from deep earth. The Arabic noun phrase والسماء ذات الرجع (was-samai zat-ir-rajyi) translated as <u>sky that rebounds</u> allude to the characteristics of a rebounding nature. The word رجع (rajyi) meaning to return or snapback is interpreted as the bouncing ability of the sky. It can be argued that this is also equally in favour of water cycle returning rain leading to gross reading error with earlier interpretation. But as we have seen in elucidation of the second sky, context found in verses dictates nature of the sky and its functionality. The salient effect of rebounding in this case is its instantaneous nature of return of exact elements which is not entirely true in case of rain held up for substantial intervals changing states of matter. And as found in the above set of verses, it is linked to the property of the earth to fissure. The Quran's deciphering of this relation between this sky layer and fissures in the earth alludes to Ionosphere's potential role in earthquake.[32] Ionosphere, structurally forms around thermosphere and in the lower regions of exosphere. This property of ionized sky is used to reflect high frequency, shorter wavelength radio waves used in

---

[32] Geoscience letters, an official Journal of the Asia Oceania Geosciences Society. A report from the Ionosphere Precursor Study Group.
Title search: 'Modifications of the Ionosphere prior to large earthquakes (EQs) on Geoscienceletters.springeropen.com.

communications. Featuring Ionosphere as the "bouncer" of a band of transmitted signals back to ground, verses 86:11-12 clearly references Ionosphere.

It is known that the Ionospheric sky besides helping in earth bound long distance radio propagation and partaking in sourcing of plasma via ionization by solar radiation; it plays important part in atmospheric electricity. Ionosphere is principled for its unique role although complex to understand with current science. Though water cycle is a part of what is returned by the shared property of the sky, it is localised within the nearby troposphere and should not be confused with Ionosphere's unique bouncing ability. Just like the other layers of skies which are geared for various aspects of protection owing to their unique abilities, so is Ionosphere—uniquely featured for its rebounding mechanism. Also, during an earthquake Ionosphere facilitates communications to various earthly beings as brought to notice by the unnerving of animals prior to earthquakes.[33]

> Quran 65:12 ...(Allah's) command descends through between them (skies) so you may know that Allah, is over all things competent and that, Allah, has encompassed all things in knowledge.

Here in this verse it is pointed that orders from Allah reach the earth through essentials in the earthly skies. It is up to us now, I mean the scientific community to bear the burden of pertinent research to unveil exact roles played by Ionosphere in the phenomenon of earthquakes.

---

[33] Wikipedia: Among the EQs prediction methods, animal behaviour is classified as a precursor to EQs.

1.6   SIXTH SKY

The collection of charged particles gathered in a couple of belts by earth's magnetic field are famously known as Radiation belts, for schematic representation see **Figure 2** and **Figure 5**. They are very flexible and can wax and wane in response to incoming energy from the sun. But they make a strong barrier capable of restricting high energy ultrafast electrons from the sun, which is a remarkable feature of these belts. This explains Radiation belts also known as Van Allen belts for characteristics of absorbing solar shocks by warding off high energy electrons from entering the near-earth space.

> Quran 21:30 Do not the disbelievers see that the Skies and the Earth were a joined entity and We (Allah) separated them and made from water every living thing? Then will they not believe?
>
> 31. And We made within the earth firm mountains, lest it should move with them, and We made therein (mountain) passes (as) roads that they might be guided.
>
> 32. And We made the sky a <u>protected ceiling</u>, but they, from its signs, are turning away.

The Arabic phrase in verse 21:32 سقفا محفوظا (saqhafann mahfuzaa) means a secure roof translated as <u>protected ceiling</u>. It is a detail told in a specific context, as the lastly constructed vault for tight protection. The Plasmaspheric raised enclosure of Heterosphere والسقف المرفوع (wa-saqhafi al-marfuaā) in verse 52:5 discussed in section 1.2 shares common attribute of a roof (saqhaf) with (saqhafann mahfuzaa) shown in verse 21:32 here. The difference between them is, one is a raised enclosure of Plasmaspheric Heterosphere and the other is a protected ceiling of Radiation belts. Both are different from بناء (binaa) discussed in section 1.4 for detailing fourth sky which meant a structure made of elemental units of Ozone. The set of verses outlined above has few outstanding points to note:

1. The primordial earth mass separated into earth and its seven skies as indicated by *Quran 21:30 "Do not the disbelievers see that the Skies and the Earth were a joined entity and We (Allah) separated them..."* First Magnetosphere from within the earth was separated to coalesce with the smoke to form the earliest skies along with gravity that by now had resulted in its full potential. Sky separation was aided by the pull of

gravity on Homospheric layers relatively shifting due to earth's tectonic plates restructuring. Gravity though does not make so much of a big influence upon Heterosphere and its layers due to the charged state of plasma.

2. Water had spontaneously precipitated early on over earth's surface immediate to the realisation of the Homospheric sky. Then primordial life was harmonized as soon as water was saturated on earth's surface. This verse could easily be at fault: if not for the words of Allah, *"and made from water life"* which directly follows the separation of skies *"and We (Allah) separated them (the skies)"* in Quran verse 21:30.

3. The formation of atmosphere was a prerequisite for water to condense and to saturate on earth's layer. If not for the separation of Magnetospheric sky from within the earth, the material of atmosphere that which was smoke would have gradually disappeared into vacuum due to solar activity and any water content would have boiled away due to absence of pressure gradient because it is the atmospheric pressure that maintains liquid sate of water on earth.

4. Ionosphere as part of initial Heterospheric sky and many other constituents required for development were later continually organised. Then towards the later part of the second phase earth building, refurbishing of earth's crust that had assumed was close to completion. Per the Quran all resources were being measured. The mountains were continually formed *Quran 21:31 "And We made within the earth firm mountains"* and rivers were streamed for purposes of sustaining life for earthly inhabitants.

    Quran 41:10 *"And He placed in the earth firmly set mountains to a high altitude, and He blessed it and determined in it its sustenance (for creatures) in four days- in accordance with (the needs of) those who seek (Sustenance)."*

    And Quran 13:3 *"And it is He (Allah) who spread out the earth and made firm mountains and (flowing) rivers: and fruit of every kind He made in pairs, two and two. He draws the night covering day. Behold, verily in these things are evidences for people of reflection."*

5. Further, three other skies were built in the second phase of sky building which is the final day of total creation time. And also, all seven skies were perfected on this day. Ozone was structured within Homosphere. Awhile

by Allah's Will, more complex life forms were advanced. As the completion of other two Heterospheric skies was realised via sourcing of plasma from lower atmosphere and Ionosphere leading to construction & perfection of Plasmasphere along with perfecting all other skies which were realised on the third day of the creation time. Thus, Heterosphere was geared for the last sealant.

6. Finally, a vault for protection indicative of *Quran 21:32 "And We made the sky a protected ceiling, but they, from its signs, are turning away"* was wound above after all other design activities on the earth and its other skies had come into formation. Tellingly, the context in this verse clearly drives the point home in favour of the Radiation belts as the final ceiling needed to secure many essential skies and the earth's space itself. Thus, seven skies were engineered for their purposes as discussed.

Excerpt, Protective shield attributed by science community: The protective shield, data gathered by Van Allen probes show that the radiation belts shield earth from high-energy particles. "The barrier for the ultrafast electrons is a remarkable feature of the belts," study lead author Dan Baker, of the University of Colorado in Boulder- nasa.gov.

Therefore, the Grand Quran concludes Radiation belts as a unique sky layer, one out of seven deemed as essential for earth's fulfilment of sustaining life for its designed purposes. This marvellous step by step creation order coming from a 7th century scripture speaks for God, it soars over and above our current human achievement which fails to unravel details of this complex event, in understanding the threads of creation sequences of planet earth and its functional build of 7 skies. Aren't these tokens enough for a considerate people to believe? At least by now it beckons our senses to credit the 7th century Quran for its accuracy in an area of knowledge where modern observation hasn't exacted clear truth. As we can see, science facts are remarkably stone scripted and unmistakably engraved in the scripture communicated by God.

1.7 SEVENTH SKY

Gravity of a planetary body besides having retention effects on the atmospheric particles is that which helped atmosphere to structure and stratify. It locks up earth with its natural satellite Moon. And further anchors it with the **saddle** known as the Sun—the **load bearer**.[34] As discussed in chapter 2 ahead, the sun, the earth and the moon system are arranged in mutually dependent satellite skies also known as the sky of gravity being the inherent and proportionate result of an astronomical body's mass. The satellite sky of influence is positioned here as the seventh sky to be discussed in the light of the Quran.

> Quran 25:61 Blessed! is He (Allah) who has placed in the sky constellations and placed therein (in a starry grid) a **saddle** (Sun) and a shining Moon.

The detailed explanation of this verse is made in chapter 2. Further in verses 55:7-9, it is clearly elucidated that the satellite sky is the force and balance of gravity launched in spatial co-ordinates. The principle demonstration of faith in God is to stand firm on justice. Thus, believers are entrusted to be just to emulate justice for everyone. Predicated on this principle God has pointed to the power in the satellite sky for its global dispensation.

> Quran 55:7 And the **sky** He (Allah) has raised and imposed the **balance**.
>
> 8. (Such) That you won't trespass in the **balance**.
>
> 9. And that you establish weight dispensing justice and that you lose not on the use of the **balance**.

The Arabic word الميزان (al-mizaan) translated as the **balance** in this context is impeccably infusing the **sky** with the characteristics of gravitation in it. The key verse 55:8 specifies the technical detail that this sky is such that you won't be able to deviate from its inherent balance.[35] In explaining the context further, the

---

[34] This concept is explained in chapter 2.

[35] Satellites launched in earth's orbits work on similar principles to that of Moon that they are in continuous orbit around earth. Orbital periods vary depending on distances of respective orbits from earth in altitude.

word بالقسط (bil-qhist) meaning -straight justice- poses a jussive mood in the verse.[36] Which is necessarily then harmonized with the presence of a command phrase—**to establish**. It implies faithfuls be able to organise justice by putting to use the balance of satellite sky. The Quran's inference is to point out effort in that direction; such that faithfuls lose not the utility of gravitational sky from use in the greater world. Such a futuristic idea that might and reach of power can be established in the satellite sky is a detail clearly conveyed 14 centuries ago. Thus, it had to be from the Lord of the worlds, giver of intelligence and capacity—a ray of light—for seekers of truth in search of "evidence" such that they may attain unfailing faith in Allah. The satellite sky which enables capacity for justice to be dispensed to nations of the world is a mind-blowing -science prophecy-. To concrete understanding of this alluded fact; let us reflect over few more verses where God elucidates more on uses of this sky for space endeavours; therefore, space travel is a given info in 7th century knowledge, which is now making present-day achievements for humanity.

> Quran 55:33-36 O communities of jinn and men- If you can pass to the boundary regions of the skies and of earth, then pass. You will not so pass except: by permission (of Allah). So, which of the favours of your Lord would you deny? Upon you be sent **space objects** of a fire brand and of **smoke**. And you won't be defended. So, which of the favours of your Lord would you deny?

Detailed explanation of concepts talked in this set of verses 55:33-36 is discussed in section 7.2. This clear exposition that men and jinn-beings would endeavour to reach the boundaries of skies couldn't be mere hallucinations of Prophet Muhammad in the wilderness of the desert. Information such as this is a neck-bending signature to any recipients free of bias.

The Arabic word شواظٌ (shuwaz) translated as **space objects** made of a fire brand and of smoke is yet another miracle of the Quran in this context. Comets and

---

[36] Jussive mood: The jussive is a grammatical mood of verbs for issuing orders, commanding, or exhorting (within a subjunctive framework).

active asteroids that enflame nearing star astrospheres[37] with sporadic tail effects vaporise ice loads of surface material, this is descriptive of the word شواظ (shuwaz). Burning of these materials results in the formation of a lengthy tail glowing halo. This elongated tail is caused by the radiation pressure and stellar wind of star astrospheres such as the sun's heliosphere. Therewith ice, dust, rocky materials, and organic compounds like methane and ammonia develop a fuzzy coma off the space object descriptive of the word ونحاسّ (wa-nuhas) referred to in this verse, translated as of **smoke**, made from ominous gases.

There can be no better explicit introduction to the phenomenon of asteroid and comet behaviour in a seventh century scripture! It is not fathomable for a desert peasant to foresee phenomenon like these amidst the harsh tribal life, concoct a theory, which coincidentally is not contradicting with his volume of rest of the work, and which also coincidentally does not contradict with established sciences. Far from this, it is found to be describing many space objects accurately. Thus, شواظّ (shuwaz) is descriptive of comets and active asteroids that are on defined trajectories. These were not perceived as distinguished objects in seventh century know-how of space bodies but is told in the Quran as prominent constituents of the satellite sky. These objects primarily are on reconnaissance as told in the words, *"Upon you be sent space objects of a fire brand and of smoke. And you won't be defended."* With their highly skewed elliptical orbits around the sun, Allah warns that such space objects are repellents and would leave us and the jinns defenceless when crossovers to sun bound listening stations are attempted across earth's exteriors—mysterious indeed. Elucidation of this in surah 55 is also an eye-opening revelation showing the accuracy of the Quran's compilation as the concept of satellite sky is brought forward in this same surah.[38] Thus, breaking

---

[37] Star astrosphere is a phrase used in this work akin to Solar heliosphere where the comets and other space objects in sphere of star's influence potentially are set ablaze due to radiation pressure coming of the star.

[38] Arabic: Surah, in English means a chapter in the Quran. However, Surah has a unique connotation which the English word won't capture. Surah is like a fortress. Circumstantial revelations of the Quran, as and when required were revealed, they weren't regular of chapters. But Gabriel taught Prophet Muhammad to distribute them into 114 fortresses known as Surahs. Each Surah therefore brings about its own perspective of imparting meanings.

all records of contemporary knowledge, a finding in seventh century, specifically detailed is found to be clearly at odds.

In yet another miraculous revelation, the Quran employs a different Arabic word شهاب (shihaab) for Meteors, used to describe a more specific case of a chase-phenomenon: *Quran 72:9 "And that we used to sit therein in stations for listening. But whoever listens now finds a meteor lying in wait for him."* Meteor is another participant of the satellite sky, these hop erratically and strike mostly in the upper regions of earth's atmosphere and at times descend to earth in event of chase. Their purposes are discussed in detail in chapter 7. Break-away fragments from comets and asteroids or small sized asteroids are known to burn up in the upper atmosphere as Meteoroids. But exhuming of space objects such as comets and active asteroids nearing star goldilocks zones is only newly known since observation of Halley's comet filmed by European and Russian space probes in 1986.[39] see **Figure 6**.

Being the very first sky of the embryonic earth mass seen at its full potential proportional to its size and density, satellite sky's realisation marks the end of the first phase of earth's formation and the beginning of second phase. Gravity is not same everywhere on earth, but it varies because of the planet's shape and topography. Earth's near spherical shape and its uneven crust finally defines gravity of our earth; it is thought to be weaker at the equator and higher at altitudes further from earth's centre. Gravity along with other skies was thus perfected to maturity mainly by the geological processes on the final day which is the ending of the total creation time of six days period of earth building: Quran 42:12 *"Thus, He completed them as seven Skies in two-days and engineered for each Sky its role."* Thus, the Quran concludes the gravity centred earth-bound satellite sky as a unique sky with its various participants seen so far.

---

[39] Look it up on Wikipedia for Halley's comet- Orbit and origin: The perihelion, the point in the comet's orbit when it is nearest the Sun, is just 0.6 AU. This is between the orbits of Mercury and Venus. Its aphelion, or farthest distance from the Sun, is 35 AU (roughly the distance of Pluto). There are many such space objects in very mysterious elliptical orbits around the sun stretching across planetary bodies.

1.8 CLASSIFICATION OF SKY CONSTRUCTION AND SUB-SPHERES

The total creation time of all earthly skies is a two-days period split into two different phases. In its first phase the sky of gravity at its full-potential, Magnetosphere, Homosphere and Ionosphere had concurred. Next, the second phase of sky building was organised with best indicators collected from the Quran, in its last day of the four-days period. From evidences the Ozone, Plasmasphere and Radiation belts were completed in the final day of the six-days period. The precursors that classify are:

1. Gravity integrally as part of the first phase in sky building as seen discussed in section 1.7. Then a couple of archaic firmaments (Magnetosphere and Homosphere) in this phase were built in the shadow of verse 41:11 as discussed in sections 1.1 and 1.3.

2. Additionally, Ionospheric formation as part of the first phase in sky building is due to its context with the earth's fissures. In the very first day of second phase of earth building; the tectonic plates of the earth were set into dynamic convulsions. Crust plates then gradually began to slide over the molten interiors to form lands as we know them today marked by mountain ranges at their interfacing boundaries. A phenomenon which lasted until the second phase completion. This setting of the calamitous convolutions of the earth's crust to fissure is tied with the Ionospheric sky. Therefore, due to this contextual inference it is correct to infer Ionospheres formation as part of first phase of sky building indicative of verses discussed in section 1.5.

3. Ozone into the second phase of sky building is due to the context implying findings of early life (microbial forms of life such as Cyanobacteria & Plankton) playing a part in oxygenating early atmosphere, and then the later sequential progression of non-sentient life forms needing cover and subsequently earth's surface becoming a couch for them. And as the sky of canopy—Ozone was structured most complex and sentient life forms such as higher flora and fauna were fashioned in the cover of Ozone indicative of verses discussed in section 1.4.[40]

---

[40] Wikipedia: Cyanobacteria, also known as Blue-green bacteria, is a phylum of bacteria that obtain their energy through photosynthesis and can produce oxygen.

4. The Plasmasphere and Radiation belts into the second phase of sky building is again in the context of sweltering oceans on earth's surface and for the Ionospheric sky as a prerequisite for the supply of plasma indicative of verses discussed in sections 1.2, 1.5 and 1.6.

Scientific understanding also imparts categorization of sub-spheres within atmosphere based upon temperature and pressure gradients. Troposphere, stratosphere, mesosphere, thermosphere and exosphere are how it is broadly layered in science diagrams. However, the Quran's classification is mainly based on functionality unlike the temperature or pressure density profiles in use by the scientists. The Quran does not reckon these five layers as distinct but classifies them in two principle layers. Homosphere, where there is homogenous mixture of gases and Heterosphere, where the gases are rarefied and stratified. Then on the Quran reckons Ozone, Ionosphere, Radiation belts, Magnetosphere and full-potential of Gravity as the principal skies due to their unique functionalities. The Homosphere comprises of Ozone layer as the protective gear. While the Heterosphere comprises of Ionosphere and Radiation belts. The architecture of the skies is akin to being overlapped. Both Homosphere and Heterosphere are encompassed within the overarching Magnetosphere which shields them from the destructive space weather. Magnetosphere being a distinct layer for its salient function. Therefore, the Quran is clearly found reminding us of the exclusive favour of Allah for providing us with the astounding seven earthly skies – a gift from Allah the Creator of the skies.

Also, Phytoplankton at the base of the food chain in oceanic life absorb energy from the Sun and nutrients from the water to produce their own food. In the process of photosynthesis, phytoplankton release molecular oxygen (O_2) into the water. It is estimated that about 50% of the world's oxygen is produced via phytoplankton photosynthesis. The rest is produced via photosynthesis on land by plants.

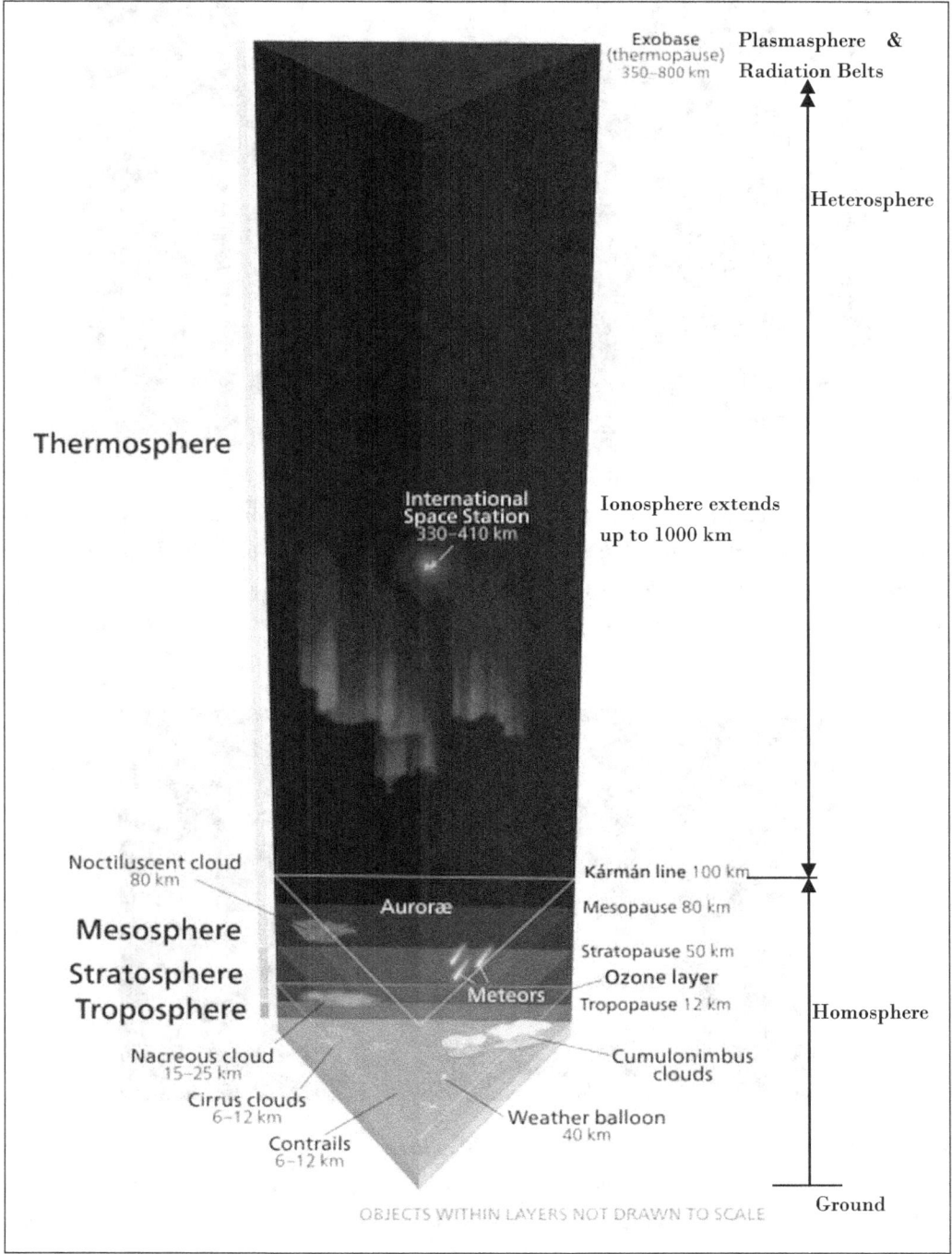

Figure 1 *Earth's atmosphere.*
Credits: Kelvinsong; Image sourced: Wikimedia commons.

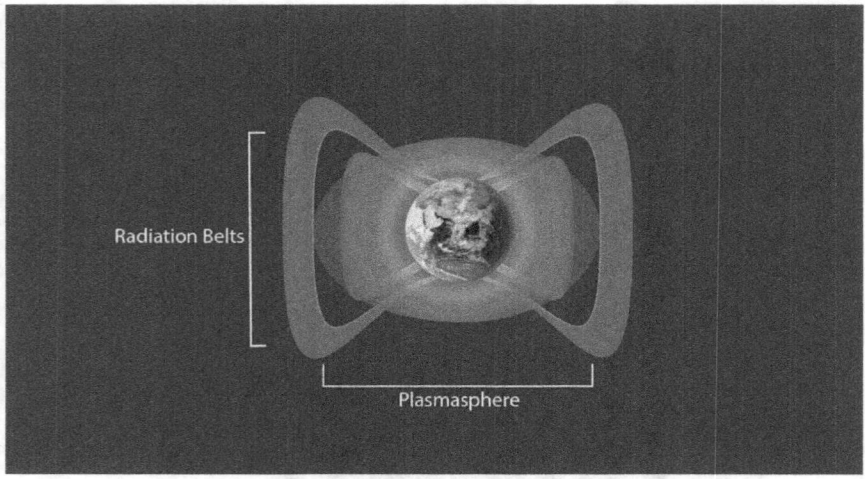

Figure 2 *Heterosphere (Plasmasphere & Radiation belts).*
A cloud of cold, charged gas around earth called the Plasmasphere and seen here in grey oval, interacts with the particles in earth's radiation belts also known as Van Allen belts—shown in a couple of grey ribbons— to create an impenetrable barrier that blocks the fastest electrons from moving in closer to our planet. Both features are neatly imbedded within Magnetosphere shown in **Figure 3** *below. Credit: NASA/Goddard.*

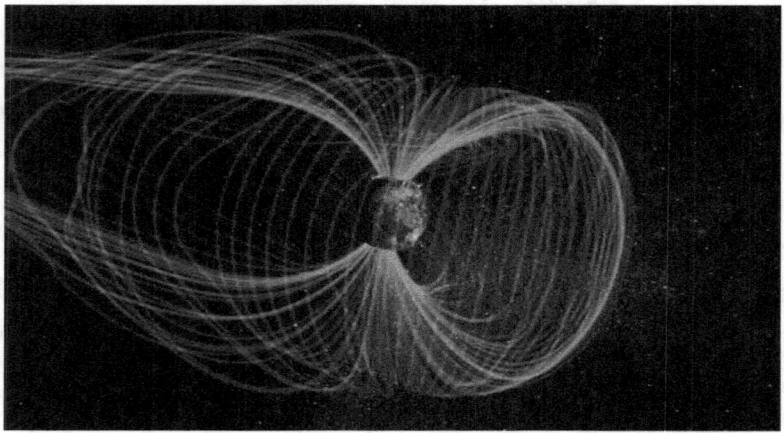

Figure 3 *Earth's Magnetosphere- Nightside extension.*
Day side is compressed to 10 earth radii (65,000 kms), night side is extended to 1000 earth radii; Earth is surrounded by a giant magnetic bubble called the Magnetosphere. Over six years in space, five spacecrafts from the THEMIS mission have helped map out this area and improve our ability to predict dynamic space weather events – events that at their worst can impact satellites in space. Credit: NASA.

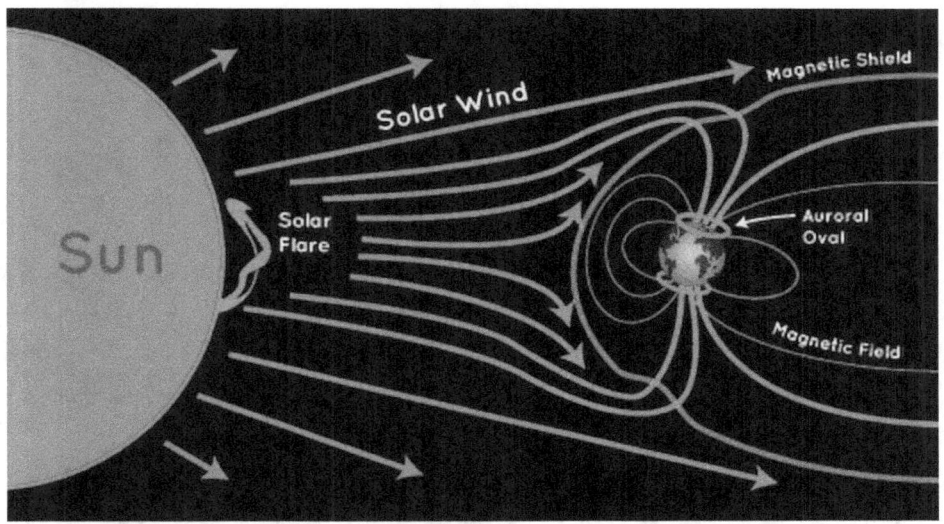

Figure 4 *Solar flare, Tunnelling Magnetosphere.*

*Illustration Image- Solar flare, tunnelling through the cusps at magnetic poles resulting in Auroral ovals. Credit: spaceplace.nasa.gov-aurora. See Aurora in **Figure 35**. Credit: NASA.*

Figure 5 *Heterosphere protecting earth.*

In June 2015, geomagnetic storm injected electrons into the radiation belts. The Magnetic Electron and Ion Spectrometer (MagEIS) instrument aboard NASA's Van Allen Probes gave scientists a new perspective on the belts' reaction. "Light grey colours indicate higher numbers of electrons," according to NASA's Goddard Space Flight Centre. Credit: NASA's Goddard Space Flight Centre/Tom Bridgman.

Figure 6 *Halley's Comet.*

On 8 March 1986, picture of a short period comet, in inner regions of solar system; Image sourced: Wikimedia commons. Credit: NASA.

2 SOLAR SYSTEM

*Light is shown to expose the drivels of rebel philosophers; chattels of
Epicurus, of any worth were the pangs of ancient Greeks. Their
tricks busted, colosseums deserted. So, why must one seek to lust
with pantheon philosophy? On the verge to lose the tressures by
reclining to watch gamers of the old!*

Heliocentric design in which the earth and planets circumambulate around
the sun is hinted to have been proposed as early as third century BC by
Aristarchus of Samos, a Greek Astronomer. But it was only widely accepted after
nearly two millennia towards transiting of the medieval world into the
renaissance of our premodern history—a century later—after Nicolaus
Copernicus had proposed heliocentric system in 1543. However, relentless
attributions favouring Greek polymaths for foreknowledge regarding science
matters albeit half-baked has dissolved the element of surprise coming from the
scriptural findings among modern cynical communities. [41] Despite Greek
polymaths' discrepancies as many as they are profuse with insufficient detailing
concerning the reality of astronomical objects their material filled with fallacies
outwardly serves well for those weary of worshipping Allah to embark on
confronting the scriptural stimulus. Concerning the creation in the Cosmos and
in origins of life, scriptural evidence presents significant impetus to dispel

---

[41] Cynical communities are referenced here as doubters in divine communications
from the Ever-Living God.

disbelief. Arguing by elegant proof found in scriptures; the Quran references its source to a non-human, nothing like it and none other than the only master creator—Allah. In among fundamentals of astronomy in contrast to literature produced by heretic-astronomers is: the sun cruises in its orbit in celestial space as told in the Quran 14 centuries ago. Which neither the Greek polymaths nor the medieval researchers had any near idea about except for observations assisted via modern instruments. Even when growing up in as recently as in the late 1980's I had in my school discrepant lessons that the sun was a stationary object and planets orbited around it. Somewhere between these times the Quranic narrative was abjectly neglected. And its exegeses by people such as (Arabic: Al-Ghazali) known in the Western medieval-world as the learned Algazelus or Algazel (1058 – 1111 CE) a Persian theologian, jurist, philosopher and mystic stood forgotten. With the Quran if you could effort to sort out the proof for yourself; it will help you realise its claim for its being miraculous in deciphering scientific information. And much more to prove that it is indeed from the All-Knowing, Almighty Lord—Allah, the giver of life, love and peace.

The Quran was revealed in 610 A.D. to Prophet Muhammad over a period of twenty-three years. Alongside its historical narrative surrounding Arabic tribes, it is found answering most wiles founded in various societies opposed to belief in the Creator and the Grantor of life. It calls itself as the definite proof from Allah. And so, you will find in it pages which seekers of truth are requested to examine fairly. Among its techniques for proofing to substantiate its claim for being Allah's word is the information given concerning minutiae involved in His creations in simple language such that even commons understand. A repeated reflection of its narrative helps dispel myths surrounding many aspects of human thought and any ensuing misunderstandings that may develop due to our lack of knowledge and reasoning as we begin insufficiently with its narrative.

Prophet Abraham (full story covered in chapter 3) regarded as the father of Prophets, had preceded Greek grandeur. To him Allah had taught the knowledge of the government in the universe.[42] Since then, time and again Allah, the creator rolled it down to mankind through divinely revealed books. It is an old-mishap

---

[42] Quran 6:75 And thus did We (Allah) show Abraham the government of the skies and the earth that he would be among the certain (in faith).

that religions post their splendorous deliverance of guidance among humanities during tenure of respective Prophets; turn out to become mere rituals amongst foolish peoples' knowhow after that period. One among several possible reasons for such straying is the jinns. Jinns outlive human life-spans. At least their chief in rebellion- Satan clearly lives a given life until doomsday told in Quran 17:61-65 which he negotiated for his eternal fall to deceive humans.[43] Thus, Satan and other disbelieving jinns in his cadre of troops with long-years of experience make sure they trip humans because of our insincere selves and not trying to be true to God. It has been a classic signature of all religions since time immemorial that post their dawn they have fallen under unbashful ritual deliverers. Satan has carefully crafted a standard of instruction which his troops follow to stray the foolish ones among us. Yet, even after Allah has asserted many revivals through a series of Prophets to better equip us, we fail to secure our interests with Almighty Lord.

---

[43] Quran 17:62 (Satan) said, "Do You (Allah) see this one (Adam) whom You have honoured above me? If You delay me until the Day of Resurrection, I will surely destroy his descendants, except for a few."

2.1 CRUISING EARTH- MERGING OF NIGHT AND DAY

> Quran 39:5 He (Allah) created the skies and earth in truth. **Wrapping** the night over the day and wrapping the day over the night and has subjected the sun and the moon. All speeding (courses) for an estimated term. Unquestionably, He is the exalted in might, the immense forgiver.

Here in this verse, the root word in Arabic كور (kawara) translated as **wrapping** means a ball, (In Google translator: type in the Arabic letters as shown to verify the meaning). Ball, in a flight takes an amount of spin. Exactly, this is the connotation implied with the use of a verb form يكور (yukawwiru) translated **wrapping**. The combination of planet earth and its skies is given to behave ball-like in an orbit with an orbital velocity and axial spin. Wrapping day into night and vice versa is a clear evidence for a spherical earth. While the sun and the moon are also subject to run spatial courses per verse 39:5, this verse clearly explains that mentioned bodies: i.e. the earth along with its skies, the sun and moon are all in their respective orbitals covering courses. This verse is certainly not indicative of a stationary earth. Or of a stationary sun as was supposed by numerous observers until lately. Many verses such as this dispel the myths of astronomical bodies including stars indicative of verses 81:15-16 (discussed in section 7.6); as were supposedly fixed objects in the sky. This verse 39:5 quoted here elegantly negates historical human guess work on the function of astronomical objects and the hypothetical model of geocentric system alleged by neo-learners who critique Islam on false tokens.

The word **wrapping** in the translation is not fully representative of the Arabic verb form used. In fact, collecting a suitable replacement isn't easy coming. Even the word **rotates**, or spin do not come close to explaining the Arabic form used. Because any flat object or odd shape about its centre of gravity-axis can be rotated and given a spin. But by wrapping, the Quran favourably elucidates this physical phenomenon with the use of another Arabic root ولج (walaja) which means to penetrate or to enter or to merge. Shown below are a couple of verses cited from two different chapters:

> Quran 31:29 Do you not see that Allah causes the night to **enter** the day and causes the day to **enter** the night and has subjected

the sun and the moon, all cruising for an estimated term, and that Allah, with whatever you do, is Acquainted?

Quran 35:13 He causes the night to **enter** the day, and He causes the day to **enter** the night and has subjected the sun and the moon. All cruising for an estimated term…

The Arabic verb form يولج (yuliju) occurring in the above quotations is explanatory of leading the skies of earth into darkness. The technical connotation is that atmospheric sky of earth is **led** into the zone of darkness, unable to refract light after being **led** into the night-side. Therefore, to penetrate, **enter** or merge, turning the splendour of the day into night which is by the planetary spin as earlier explained in this section. Both verbs forms يكور (yukawwiru) meaning wrapping and يولج (yuliju) entering or merging do not support the idea of a stationary earth. Neither does the Quran anywhere else imply to such a nonsensical assertion alleged by its critics. Further, all verse endings cited here emphasise that all bodies: i.e. the earth with its skies, the sun and moon are running orbital courses (cruising). This is thus a clear enough verdict to close discussion on this subject.

2.2 MOON REFLECTING SUN'S LIGHT

Quran 91:1 By the sun and its brightness.

2. And (by) the moon when it **reflects** it.

3. And (by) the day when it displays it.

4. And (by) the night when it covers it.

Away from false allegations the Quran nowhere alludes to common orbits of sun and moon as alleged by critics. The Quran references the world view from a human reference frame for obvious reasons of aiding relative understanding for many generations of people who couldn't visualise a bigger picture as we can today. Since in prose Quranic verses couple the mention of sun and moon as astronomical objects orbiting, like in verses seen in the previous section: *"and has subjected the sun and the moon"*, critics grow suggestive of attributing common orbits to them which is clearly an erroneous reading.

Verse 91:1 "By the sun and its brightness." literally performs a two-noun sentence (sun & brightness) with two prepositions (by & and) and a pronoun (its) in its Arabic syntax. With the pronoun ending (its) suffixed in Arabic to the second coming noun, it is translated as **its brightness**; which alludes to its first noun the sun. Therefore, the sun's brightness is praised. The next verses from 91:2-4 in Arabic rhyme, each present verb endings with a pronoun calling upon a noun from verse 1 i.e. reflects it (2), displays it (3), covers it (4). Regardless of which noun in verse 1 (i.e. sun or its brightness) is referred by these subsequent three pronouns down the rhyme, the moon in verse 2 does not tuck itself into the orbit of sun to follow it. The traditional Quran translations have assisted this erroneous reading for their use of the word "follow" instead of using the word "to reflect it" for translating تلاها (talaha) which means showing brightness again, in short "to reflect it." In Arabic, to follow within the footsteps of a lead, there is a variant word. For such a connotation تبع (tabaā) meaning to follow or go after is used. Ample use of this in relevant connation can be found in the Quran like in verses 37:10 and 15:18 describing meteors in pursuit of rebellious jinns, but not here in 91:2.[44] Here, the Arabic verb تلى (tala) mostly translated as **followed by** is used, having a root in Arabic تلو (talaw) meaning "again." Which if reworded the verse would read, the moon again shows the sun's brightness; instead of reading the moon follows its brightness. Regardless of traditional translations reasoning through context clarifies that the second noun "brightness" from verse 1 is brought forward in verses 2-4. Upon little reflection, it further becomes clear with verses 3-4 that sun's brightness is talked of as being displayed during day time and evaded away during night time.

> Quran 25:61 Blessed is He who has placed in the sky constellations and placed therein a lamp and a **shining** moon.

Here in this verse the active participle منيرا (muneera) translated as **shining** is in the accusative case of form 4, in Arabic grammar.[45] Its root word means the glow of light. It therefore entails that the moon is an actor or the doer, as also previously explained by the verb in verse 91:2 تلى (tala) meaning to show sun's brightness or reflect brightness from the sun, shining the glow of light thereof; rightly describes the moon by an adjectival noun منيرا (muneera) meaning lightsome. To illustrate this difference further in clarifying detail: the characteristic nouns for the sun and the moon are clearly two different substantives and are never exchanged or mutually shared in the Quran. Sun is always the سراجا (siraja: a loaded saddle), وهاجا (wahhaja: a burning lantern) and ضياء (dhiyaa: a shining light source) whereas the moon is نورا (nur: glow of light) and منيرا (muneera: lightsome). By this demarcation it is very clear that the moon shines forth light or brightness incident from sun. In this sense, it is traditionally translated in the verse as followed by the sun's light or brightness. However, in scientific etymology it must be precisely technical and therefore newer translations of 91:2 must read as "reflects it" instead of "followed by it." Thus, the reworded verses are as follows:

> Quran 91:1 By the sun and its brightness.

---

[44] Discussion on celestial following as in a pursuit by the root تبع (tabaā) is made in section 7.13.

[45] Internet search result: The active participle is essentially an adjective closely related in meaning to the meaning of the verb. However, active participles are often also used as nouns.

2. And (by) the moon when it reflects it (sun's brightness).

3. And (by) the day when it displays it (sun's brightness).

4. And (by) the night when it covers it (sun's brightness).

Needing no further elucidation, the Quran is seen to conclude that moon isn't a self-luminous body like the sun. Moon therefore has a different role to play for which it traces sun's brightness in phases showing it magnificently to the dwellers on earth for the purposes to trace the count of time as discussed in the next section. In fact, the Quran points out that even the sun is used for celestial time keeping, acknowledging the benefits of both seasonal and lunar calendar's respective entitlements: *Quran 6:96 "(Allah is) the cleaver of daybreak and has made the night for rest and the sun and moon for calculation. That is the determination of the Exalted in Might, the Knowing."*

2.3 PHASES OF MOON

> Quran 2:189 They ask you (O Muhammad) about crescents.
> Say, "They are measurements of time for the people and for
> **Hajj**." And it is not righteousness to enter houses from the
> back, but righteousness is (in) one who fears Allah. And enter
> houses from their doors. And fear Allah that you may succeed.

Naturally owing to revelations of the Quran and its enlightenment as a
pristine source of guidance, the Arabic tribes once submerged in superstition
could now query objectively. The wrong cultural practices interwoven with
formalities of using their back doors to enter homes at the sighting of crescents
was dispelled as not a righteous act anyway. Allah clarified that new moons are
for measurements of time for people and for endeavouring for **Hajj**—a pilgrimage
to Makkah, to mark the symbols of Allah's favours upon mankind in the footsteps
of Abraham, Hagar and Ishmael's profoundly sought sacrifices.[46]

A leap day in solar leap year-cycle is essential to fix the orbital lag of the planet
earth revolving around the sun. But lunar calendar is free from such a blemish.
New moon in crescents return to the horizon once every lunar month to mark its
beginning. Given that the number of months in a year is twelve as reckoned by
Almighty since the day He created the planet earth and its skies. Therefore, the
need to alter time to correct the number of days is unwanted in lunar
measurements: *Quran 9:36 "Indeed, the number of months with Allah is twelve
(lunar) months in the register of Allah (from) the day He created the skies and the
earth; of these, four are sacred. This is the correct religion, so do not wrong yourselves
during them…"*, inference of since this event of creation is also evidence for the
creation time of planet earth and its skies of belonging to a different six-day
periods and not 24 hrs. Moon was made to orbit around earth in consistent
intervals generating monthly calendars of regular 24 hr days. Solar calendar
though does help closely monitor seasons for farming benefits nevertheless
seasons are not far from obvious. If the moon were to follow sun along its orbit as
some critics of the Quran understand it—a shabby glitz—captured and refuted

---

[46] Abraham (Arabic: Ibrahim), Hagar (Arabic: Hajirah) and Ishmael's (Arabic:
Ismail).

in pervious section; then, the Quran would not have drawn attention to the moon's stages as it hasn't rightly drawn it for the sun.

> Quran 10:5 It is He (Allah) who made the sun a shining light source and the moon a glow of light and determined for it phases – that you may know the number of years and estimate (of time). Allah has not created this except in truth. He details the signs for a people who are knowing.

Both bodies the sun and moon have their respective orbits. Eclipses were a widely known phenomenon which had even surmised forms of superstition among Arab cultural lives. Even they in much ignorance would have hooted at this allegation of forcing the idea on the Quran for common orbits to sun and moon. There isn't any scope for cherry picked interpretation coming the Quran's way. The Quran is a self-explanatory book of Allah, The Wise and Knowledgeable.[47]

> Quran 36:39-40 And the moon – We have measured for it phases, until it returns like the old date stalk. It is not for the sun to trespass it, that it must **outrace** the moon. Nor does the night outstrip the day. All but in their orbits cruise.

The above verse talking of moon's returning with its final phase wherein it appears like pale yellow stalk of a date palm is indicative of its orbiting around the earth. As the verse is clear on the sun and the moon not being in common orbits for a race to **outrun** each other. But all bodies are in their respective orbits cruising as the verse ending goes—all but in their orbits cruise. It is not up to the sun to side-track with moon. Both cruise in their respective orbits. The Quran verses 36:39-40 are clear in telling the fact that they (sun and moon) are separated in respective orbits. The Arabic verb form تدرك (tudrika) means to outrace coming from the root word درك (daraka) meaning riding a horse to overtake. The word ينبغي (yambaghi) meaning to trespass complements the denial of an outracing-interpretation, that the sun is in no position to interfere with the moon for a race. Nor can darkness outstrip the day-side. The Arabic word here is سابق (sabiqh) rendering the sentence to mean that neither can night advance into the day or vice versa. But all bodies: i.e. the earth with its skies, the sun and moon cruise in

---

[47] Quran 11:1 Alif, Lam, Ra. (This Quran is) a Book whose verses are perfected and then elucidated in detail from (one Who is) Wise and Acquainted.

respective orbits, summing up a crystal-clear idea of the planet spinning in its axis with its moon phasing monthly and then all cruising in celestial spaces along the load bearing sun is an exhilarant—ancient narrative.

Such is the narrative saga of Allah, of the creation act that is extolled in these chapters.

2.4 SUN A STAR AMIDST CONSTELLATIONS

We know sun as a star in the Orion constellation. More fitting a star in the familiar solar neighbourhood. It is perhaps new knowledge that stars like our sun are equally or more massively gravitating and even so more fiercely burning. Amazingly, the Quran squarely places our sun **casually** as a unit star amidst the larger part of a patch in the starry sky. But alludes to its apparent capacity of burning like that of a lamp along with immensely gravitating spatial ordinates which help bear its load, gripping and illuminating its space with splendorous brightness.

> Quran 25:61 Blessed! is He (Allah) who has placed in the sky constellations and placed therein a **packsaddle (sun)** and a shining moon.

The noun that describes sun in verse 25:61 is سراجا (siraja) a proper synonym for sun used widely in Islamic arts and culture. It translates as lamp or a burning body. Its root سرج (saraja) means a saddle, packsaddle or a harness and therefore explains the sun as a bearer of load. Clearly, it is indicative of a heliocentric system. The reason for its connotation to popularly mean a lamp of light and not a packsaddle is because of its synonymous usage for sun enrooted in Arabic culture and in Islamic traditions. It is a term covalently used for sun without the scientific impetus to realise its gravity—acting like a saddle. Hence, correctly understood translation reads as follows:

> Quran 25:61 Blessed! is He (Allah) who has placed in the sky starry constructions and placed therein an **anchor** (sun) and a lightsome moon.

Further, as in averting confusions away this term سراجا (siraja) is critically remarked by an adjective which translates for being a lamp as follows:

> Quran 78:13 And made an **anchor-lamp.**

Here in this verse, the Quran hammers the nail to its full-depth neat and tidy. It defines the anchoring object sun سراجا (siraja) by an adjective وهاجا (wahhaja) having a root وهج (wahaja) meaning glare, glow, heat, blaze, flame or lamp. Therefore, the sun is a burning saddle. Thus, rephrased translation would read as follows:

Quran 78:13 And made a **gravitating lamp**.

Traditional translations of the Quran in this area make a poor presentation because the ability to knowledge and inquiry in those days wasn't as farfetched as it is today.

> Quran 36:37-38 And a sign for them is the night. We remove from it (the light of) day, so they are (left) in darkness. And the sun runs (its course) as is **determined** for her. That is the decree of the Majestic, the Knowledgeable.

The set of verses from chapter 36 quoted here begin publicizing a sign for disbelievers who did come to think that Allah hasn't made any effort in reaching out to them for demanding faith from them. So, it says removal of daylight plunges the planet into darkness, explained by the wrapping of day and night due to planetary spin as discussed earlier. Further, the sun's run or float in the sky which must necessarily be interpreted in terms of speed, velocity or orbital cruise is seen appraised by a verb تجري (tajri) ahead of the passive participle connecting it with a preposition لمستقر (li-mustaqhar) meaning "**determined-for**" (the sun).[48] Also, worth noting is the mistranslation of this bit. Traditionally, it is implied that sun is travelling to a resting point. But my disagreements are in-place. The Arabic verb تجري (tajri) meaning orbital cruise is a precursor and will invoke a contradiction if the passive participle لمستقر (li-mustaqhar) would mean a place of rest. Thus, the preferred translation is *the sun runs its course as is determined for her*. In Arabic syntax, this nominal sentence ends on a feminine pronoun with a preposition لها (la-ha) referring to the feminine noun the sun; indicating the determined orbital cruise of the sun. This is nothing less than a heliocentric system in cruise in celestial space.

The Quranic narrative is simple and factual. Whether the sun may orbit the centre of mass within solar system known as the barycentre or goes around the centre of mass in the interstellar neighbourhood cruising around the galactic

---

[48] Internet search result: Passive participle is a form of a verb that in some languages, such as English, can function independently as an adjective, as the past participle "baked" in- "We had some baked beans", and is used with an auxiliary verb to indicate tense, aspect, or voice, as the past participle baked in the passive sentence; The beans were baked too long.

centre, it is a loaded saddle that carries its load of planetary systems and discrete space objects. Our solar system's barycentre constantly shifts position which depends on planets' relative cruises within their orbits. The solar system's barycentre can be near the centre of the sun to being outside the surface of the sun. As the sun dominates the barycentre, it wobbles around. This wobble effect is the key arm in locating distant stars for the possibility of hosting exoplanets. The Quran's describing the sun as the bearer of load surfaces as the most accurate description. Thus, in such simple words Allah answers the allegation that nothing thus far has reached us from Him, because it has indeed reached, if we can only see!

2.5 CREATION OF SUN AND MOON

Verses concerning creation of planet earth and its skies are diversified. Uncovering those details will help understand their creation sequence with the Sun and Moon.

> Quran 25:61 Blessed is He (Allah) who has placed in the sky starry constructions and placed **therein** (in a starry grid) an anchor (Sun) and a lightsome Moon.

> Quran 71:15-16 Don't they reflect how Allah has created seven skies overlapped. And has made the Moon **amidst** as lightsome. And had made the Sun an anchor.

> Quran 78:12-13 And We (Allah) built above you seven strong (skies). And had made a gravitating lamp.

In verse 25:61 Almighty Allah says: He has made the Sun as a star **amidst** one starry grid of the many constellations in the sky. Also, the Moon is placed **within**. The word **"within"** is indicative of the Arabic preposition plus a singular feminine pronoun فیها (fi-ha) translated in the verse 25:61 as **"therein"** alluding to a star-cluster of the many found in the dominant sky space of the galaxy alluded as -sky-. Both objects the Sun and Moon are called adjacently in the verse belonging to the same sky space of starry grid i.e. solar neighbourhood, which will be referenced herein as the first cosmic or major sky (detailed in coming sections) and different from the seven earthly skies seen so far. Further, reflection of next quoted set of verses 71:15-16 tells: After Allah had created the planet earth and as its skies were drawn out—Moon was placed amidst its skies. The Arabic word for a joined preposition and a plural feminine pronoun فیهن (fi-hinna) used here translates as **"amidst"** or **"within"** alludes to earthly skies. The calling of Moon in here is syntax specific, breaking the adjacency with the sun as observed in verse 25:61 (where major Cosmic sky makes common reference to both sun and moon). Thus, the moon was placed amidst the seven skies as lightsome and then this whole system had been placed to anchor with the Sun.

Information sequenced in verses 71:15-16 is concreted in the lastly quoted set of verses 78:12-13 wherein Allah says: He built seven strong skies and had anchored this system to the gravitating lamp, the Sun. As in this set the Arabic فیهن (fi-hinna) is missing making it clear that the gravitating lamp i.e. sun is not amidst the earthly seven skies. This therefore tells the important difference in the orbits of Sun and Moon. And emphasises the Moon's proximity to earth as a body situated within its sky-space.

Moon is placed within the sky-space of the earth, orbiting around it as its satellite, showing phases to keep times of the lunar month. As planet earth after its creation completion was tuned to assume the clock of twelve lunar months [9:36], this tells Moon's functional readiness to coincide with that of the time of earth's creation completion. [49] The Moon being within the earthly skies [71:15-16] supports the narration of its creation story to have concurred sometime after the first phase of the formation of an embryonic earth-mass. This tells the Moon's origins to have come about in the second phase of the earth building activity to its full development when all earthly skies were drawn out and perfected. Most importantly as the proportional size of the earth was long realised and its full gravitational potential unleashed to entangle with its satellite Moon. Making of the Moon is rightly indicated to have synchronised in the second phase of earth building (four-day period). Prior to this happening Moon could not have been founded to grid lock with the pre-embryonic earth-mass. Therefore, the Quranic information that the Moon was placed in the earthly skies in their drawing out of the earth is an elegant deciphering of knowledge. Precisely, the Moon was fully developed to assume its orbiting courses covering monthly phases—about when all the earthly skies were realised at the end of the combined second phases of the earth and the sky building process. And as this whole system had been already placed to anchor the Sun this tells Sun's sequence of creation to predate all its load (earth, other planets and entities) within their respective orbitals surrounding it. The setting of clock and time: *Quran 9:36 "Indeed, the number of months with Allah is twelve (lunar) months in the register of Allah (from) the day He created the skies and the earth..."* as highlighted earlier is clearly a conclusive proof of creation time of six-days being of a different time period. 24 hrs-based time in fact was realised after creation completion of planet earth, its skies and the time keeper moon after having all rotations and revolutions of these bodies fine-tuned within orbits; as the verse here deduces the number of months in a year as twelve, since the day of their creation completion.

---

[49] Quran 9:36 Indeed, the number of months with Allah is twelve (lunar) months in the register of Allah (from) the day He created the skies and the earth; of these, four are sacred. This is the correct religion, so do not wrong yourselves during them...

In modern astronomy the idea of a large asteroid hitting embryonic earth and separating a large chunk out of it is strongly postulated as a precursor to birthing of moon. However, the Quran does not comment on this possibility but simply pens "and God made the moon amidst the skies of the earth" which could have been independently realised by another gravity contrail that began accreting matter for moon just as protoplanet earth began forming from the initial clay cloud that surrounded the solar nebula.

Thus, it is concurred that unambiguous exposition is found for the astronomical objects in the Quran giving not only the direction to the sequence of creation, but also their positions within the solar system. Looking at all these astronomical data I realise that if the Quran was man-made, all it had to do was make one mistake- one haywire statement about the concerned astronomical body which would throw it wide off the grid; for example, if sun and moon were to interchange places in 25:61 – It would as they say, "screw up" everything. One mistake and the object would be hurled a billion miles off its actual station of description. It is to this degree that precision of information is demanded from the scripture. As we are still in the 2nd chapter, the critics looking for that one mistake have much game left to play.

3 ANCIENT ASTRONOMY

If you could shun the heretics—for the evidence is intact in favour of young Abraham!

Sky of our world which is the solar neighbourhood is our first Cosmic sky. The Quran's idea of our world as mentioned earlier is: the sun, the moon, and the earth system as experienced from human reference frame. Then the whole lot of material load of our solar system forms the part of a protective gear along with the neighbourhood flock of stars within this sky space. This first Cosmic sky is where our solar system resides with its neighbours among the most prominent stars of our galaxy. It is also adorned with presence of remarkable astronomical bodies such as the display of fire cracking meteors, from the vast spreading necklaces of Asteroid belt, of the Kuiper belt, of the massive overshadowing Oort cloud bubble and the planetary bodies cruising in whorls around our sun with many of the brightest and most famous celestial bodies (stars and nebulae) that are neighbours to it.[50] The beauty in this artistry prominently of stars in the night sky simply lacks words for any standard of praise.

> Quran 37:6-7 Indeed, We (Allah) have adorned the near world Cosmic sky with an adornment of **bodies in whorls**—And as protection against every rebellious Satan (Jinn).

---

[50] Detailed discussion on space objects such as non-starry and non-planetary astronomical bodies is brought forward in section 7.2.

Verse 37:6 here expresses the Arabic phrase بزينة الكواكب (bi-zinati al-kawakib) translated as, the adornment of astronomical bodies in whorls; this is descriptive of first Cosmic sky as being loaded with gadgets and heavy brocades of space objects to reckon with, a rich interstellar colony with millions and millions of non-starry bodies in whorls. The Arabic word كوكب (kawkab) means an astronomical object in a whorl; it is mostly taken to mean a star or a planet or other non-planetary object. In the Quran, it is in fact a word that is used to connote all astronomical bodies: stars, planets and non-planetary objects within unique syntaxes depicting combination of stars, prominent planets and many other non-planetary bodies found in whorls. In Allah's own narrative here in this specific context of verse 37:6 it is told as inclusive of all the above descriptive of its purposes meant for protection. However, in its own style the Quran clarifies that كوكب (kawkab) should not always be taken to mean a star by narrating a scene from Abraham's story as detailed in sections ahead.

3.1 ABRAHAM- FIRST ASTRONOMER

> Quran 6:75-76 So also did We (Allah) show Abraham the power and the laws of the Universe[51] and the earth, that he might (with understanding) have certitude. When the night covered him over, He saw a kawkab: He said: "This is my Lord. "But when it set, He said: "I love not those that set."
>
> 77. And when he saw the moon rising, he said, "This is my lord." But when it set, he said, "Unless my Lord guides me, I will surely be among the people gone astray."
>
> 78. And when he saw the sun rising, he said, "This is my lord; this is greater." But when it set, he said, "O my people, indeed I am free from your abominations (of worshipping objects).
>
> 79. Indeed, I have turned my face toward He who created the skies and the earth, inclining toward truth, and I am not of those who associate objects for worship."

In the very distant historic past, young Abraham (Arabic Name-Ibrahim) was an emerging sceptic from the works of his people who were sunk deep in abominations by their choices of numerous deities for worship. Allah then was his aid, for Abraham had a determined will to employ reasoning. Thus, Allah ushered him with knowledge of the laws of the universe and that of the earth, *"So also did We (Allah) show Abraham the power and the laws of the Universe."* As Abraham grew bolder to put to test the claim of divinity, he saw a كوكب (kawkab)

---

[51] Universe is a deduced translation for the word السماوات (as-samawat). There isn't any such precedent translation in the Islamic traditions. Herein it is deduced because the Quran places the earth at the centre of a grand show unfolding with a couple of intelligent beings. Therefore, in many contexts in its narrative it calls to infuse the realm of all skies: Earthly and Cosmic in the very term السماوات (as-samawat) and then calls the planet as a separate entity. In this work, the term universe is employed to infer the realm of Allah's creation, which for sanity reasons excludes His abode which is worthy of Himself, His residence from above encompassing the universe. The gapping question of what resulted in Allah for agnostic minds will find fulfilment at the turn of –the certainty. Which is via death and resurrection, a promise upon Allah. Post resurrection our conscious self which once procrastinated earning faith having had given enough evidences; then on will only lament as then faith will have ceased from being a commodity of want. At when the nature of God-the truth, will be fully realised as—the only uncreated being.

which is a planetary body; mostly translated as a star in many of the Quran translations. It is wisely sensed, that in the history of humanity much knowledge hasn't been documented. Especially, in the wake of the Quran's deciphering stories of the previous people this is becoming obvious. In Abraham's case of early reasoning, the Quran uses a singular form "kawkab" of the plural kawakib as an astronomical body upon which Abraham focused reflection in the night sky. Understandably night sky would be showing many stars. The clear qualifying reason for using the singular Arabic word كوكب (kawkab) is brought to light here, as to how Abraham graduated learning of a planetary body. It was not a star that Abraham grew keener to focus but it was a "planet" that he locked his eyes on. Perhaps then he hadn't realised to distinguish planets from the starry beacons; but, after he had gathered much knowledge he still shunned the power of stars for any worth of his worship. In twelfth-century, Qurtubi an exegete, comments on this verse highlighting that Abraham had locked his gaze on either Jupiter or Venus.[52] And Jalalayn, a much popular but later exegete of the fourteenth-century has emphasised that Abraham had spotted Venus for reflection.[53] Both exegetes clarifying that they knew the difference between the planet bodies and stars as they did not translate the Arabic word كوكب (kawkab) for a star in this context in verse 6:76.

Venus in the night sky appears brighter than any other planet or a star at its brightest apparent magnitude of -4.9, see Figure 7 and Figure 8. Venus transits between being an evening star, visible in the sky after sunset, to become a morning star, visible from hours before sunrise and for Abraham it was like an apple caught from the numerous stars that trickled the night sky. The grand Quran tells us that the كوكب (kawkab) in this case a planetary body which Abraham, not supposing it as such; drew the attention of his people debating with them as his potential superior deity to theirs many sculpted and carved deities of mud and wood. But he noticed that this kawkab would retrieve from the drama of the night sky at regular intervals. At length after many realisations, he said to his people, "I won't consider that which desert the stage." Which very likely argues the case in favour of Venus as the alluded كوكب (kawkab) in the verse. Because, Jupiter which is next to Venus in brightness and at times also next to

---

[52] Tafseer-E-Qurtubi on archive.org; volume 8. Mushtari: Jupiter and Zahra: Venus.
[53] Tafseer-Al-Jalalayn on archive.org.

Mars but remains appeared throughout the night sky rising at dusk and setting at dawn qualifies less for the object brought forward in this verse. In the story line, Abraham strenuously applying reflections repeats similar exercises wanting to face his Lord, for he was acclaimed by otherness, not to see one facing before him, after having been raised seeing numerous idol gods of his people all his life. The next target in the night sky obviously became the Moon. After many astronomical reflections of the moon's stages now more apparently observable from Venuses, [54] Abraham seeing moon's appearances, phases of waxing & waning, extended stays during dayside, receding from the sky in part of the nights and non-attendances: he clarified to his people "Unless my Lord guides me I'll stray." And then he drifted to observe the day drama and focused on the powerfully shown countenance of the Sun, remarking that it was superior. Further, before the sun's departing saga at every dusk, his reflections shifted towards the many distant stars emerging at nightfall; clarifyingly, he again said to his people "O my people I free myself of ascribing materials for worship." Then looking at the stars, [55] he said, I'm sick! Awhile his people grew in suspicion nearly expecting Abraham to declare a new set of deities in multitude after his boredom from the immediate objects whom he saw as setting. They had thought that he had stretched too far and wide to take stars for a change in divine wonderment. But Abraham was sick in the mind from anymore of such wrongful ascribing. Although in short frames of time stars didn't appear to set but only shift, for him now many of those did not make the Almighty Lord of the worlds. He said to them he felt sick in the mind to do such a thing and that he wasn't going to reckon the stars for any such worth and, so they departed from him. By then for his many deep observations of astronomical phenomenon, immense and powerful realisations had set in, enforcing high-mindedness as he concluded—I'm sick of hoping an Almighty Lord in the artistry of the universe.

> Quran 37:84-90 Behold! he (Abraham) approached his Lord with a sound heart. Behold! he said to his father and to his people, "What is that which you worship? "Is it a falsehood-

---

[54] Venus shows phases like moon see **Figure 8**, which can be easily observed with a small telescope or binoculars. Phases of Venus on Wikipedia claims: Acute eyesight in fact helps make out the phases of Venus without any optical aid.

[55] Quran in verse 37:88 uses the proper Arabic word النجوم (an-nujum) in plural for stars.

gods other than Almighty- that you desire? "Then what is your idea about the Lord of the worlds?" Then did he cast a glance at the Stars. And he said, "I am indeed sick!" So, they turned away from him, and departed...

91-99. Then he turned to their gods and said, "will ye not eat (of the offerings before you)?... "What is the matter with you that you speak not (intelligently)?" Then did he turn upon them, striking (them) with the right hand. Then came (the worshippers) with hurried steps and faced (him). He said: "Worship you that which you have (yourselves) carved? "But Almighty has created you and your handwork!" They said, "Build him a furnace, and throw him into the blazing fire!" (This failing), they then sought a stratagem against him, but We (Allah) made them the ones most humiliated! He said: "I will go to my Lord! He will surely guide me!

Thenceforth, he laboured to know his Lord beyond the material makeup.[56] Sound in spirt he turned himself unto the Lord of the worlds. Now, deciding to conclude the debate he tackled his father and the chiefs of his tribe saying: Do you invent falsehoods as your idols—deities for worship? Then, how do you reason concerning the Lord of the worlds? They said, "We found our fathers worshipping them." Swiftly Abraham replied: you and your fathers of old have been in error. Nay God is one who initiated origins and I am a witness to this truth. This is his profound conclusion after his strenuous reflection of the observatory that God had offered him. So, Abraham faces their deities in the temple of worship, talking aloud to the idols commanding them to dare share a meal and make a discourse. Then he began with his right hand breaking them into pieces except the biggest of them, that they might turn to it. Then they brought him to answer for it. Abraham said, this was done by the biggest one, ask them if they can speak intelligently. Thus, they were overwhelmed with shame: saying, you know they do not speak! Abraham assumed "Fie upon you for worship of such things beside Almighty Lord of the worlds." Therefore, he was apprehended in criminal chains.

---

[56] Fundamental entities in this universe are made of mass i.e. matter and things without mass i.e. energy. Light is of the latter. Plasma and gas are both states of matter. But light is not. Allah is not Light himself. Albeit He draws a similitude of Him as being the Light of the universe and that of earth in Quran 24:35. Which is an allegory. For there is nothing like God as clearly told in Quran 42:11 and 112:4.

Then enquired the Lord in his land whom Allah had given power of rule. Abraham said, "My Lord is He who gives life and death." So, the ruler said, "so do I, take life and spare it from whomsoever I will." Said, Abraham: "My Lord brings the sun from the east side so can you cause it to rise from the west side?" Thus, the ruler was confounded. And, Abraham had emerged an Astronomer and a sincere, seeking truth. So then, Allah made him a Prophet and communicated tidings of this office by sending Arch Angel Gabriel, which was a blessing upon him for his acutely reasoned and unwavering conclusions of faith in the Master Creator of the universe. Then did Allah confer upon him compassionate Ishmael[57] and a brilliant Isaac[58] who fathered two great nations. Hence Allah chose Abraham's lineage to bestow exclusive favours of the office of Prophethood.

> Quran 2:258 Haven't you considered the one who disputed with Abraham concerning his Lord, because Allah had granted him power? Abraham said: "My Lord is He Who gives life and death." He said: "I give life and death." Said Abraham: "But it is Allah that causes the sun to rise from the east: Do you then bring it from the West." Thus, the disbeliever was confounded. Nor does Allah Give guidance to a people unjust.

In conclusion the Quran undoubtedly presents Abraham as the first Astronomer who in his coming of age applied immense reflections at the unfolding of the natural phenomenon before him. Therefore, he was praised by Almighty Lord Allah Who then further guided him.[59] His reflections had helped him learn Venuses phases from a greater distance and of the Moon's more apparent phases from a near distance. After his diligent observation of phases, it had manifestly explained him that they weren't self-luminous bodies. And then the Sun had shown him that the earlier bodies of his observation were in fact reflecting light of it. And that they were within the influence of it as a near lamp. And that the Stars which were more distantly located, and which did not show any signs of

---

[57] Quran 37:101 So We gave him good tidings of a forbearing boy.
[58] Quran 15:53 (Angels) said, "Fear not. Indeed, we give you good tidings of a learned boy." Also, see verse 51:28.
[59] Quran 3:191 Men who praise Allah, (all times) standing, sitting, and, lying down on their sides, while contemplate the (wonders of) creation in the universe and the earth, (With the thought): "Our Lord! not for naught did you created (all) this! Glory to Thee! Give us salvation from the penalty of the Fire.

dependence on sun, but had within them the capacity of the Sun. Further due to learning of Venuses morning and evening appearances in the sky he had grasped Earth's rotation. Deducing that this "kawkab"—Venus is not -in charge- but is regulated; it appears and draws behind stage in intervals. Also realising Venuses orbital lag like the Moon's easily traceable, Abraham had learned bodies were traversing in the sky. Further, Moon's monthly renewals of the lunar month had told him of its belongingness to the Earth, a satellite in space. And that Venus showing subtle phases for its difficulty of sighting than the more apparent phases of the Moon, this was therefore telling him Venus was located at a greater distance away from earth than moon. Also, Venus's morning and evening tidal appearances had him learn of earth's daily rotations causing sun rise/set and further Venus's common belonginess to the sun was understood. And that the brightness variation of Jupiter with respect to Mars, for Mars's more apparent orbital lags had told of their similar tale comparable to that of Venus's also as non-starry astronomical objects (kawkab-s). Further, Abraham's understanding of the realm of astronomical distances of Stars due to their markedly lesser horizon lags, all of which was probable if the near immediate bodies that drifted more and phased were a part of greater system at work traversing in space along with the Sun, under its influence. When Abraham had understood such finer details he thus presented his mastery on this subject in certitude before the seated Lord of law in authority in his land. Saying, "can you then cause the earth to show: the sun from the west", thus perplexing him. This is the meaning of his comprehension of the laws in the Universe and of the Earth which Allah had given him as asserted in verse 6:75 cited earlier in this section.

Then, as the story is continued...

> Quran 12:4 When Joseph said to his father (Jacob), "O my father, indeed I have seen (in a dream) eleven كوكبا (kawkab-ann) and the sun and the moon; I saw them prostrating for me."

Prophet Joseph (Arabic Name: Yusuf) a great grand-child of Abraham from the line of Isaac (Arabic Name: Is'haqh) fathered by Jacob (Arabic Name: Yaqhub) was another high-IQ scholarship kid grasping the knowledge from the heritage learnings of Prophet Abraham. Joseph's dream talked in the Quran could be read to allude to eleven most appealing as planet bodies in our solar system with the honour of twelfth planet Earth alluding to himself in his dreamed allegory. Featuring his dream, the Earth is the jewel in the solar system for which the Sun and the Moon make prominent bodies showing light as a parent and help. Thus, it was prophesised to him that he will represent the office of Allah in the family over his eleven brothers who would be led by him. In traditions the Arabic word كوكب (kawkab) has been given a normative translation of stars to describe of his dream featuring his brothers. Eleven كوكبا (kawkab-ann) mentioned in this verse could be eleven Planets instead of Stars because kawkab simply means an astronomical object. This is of course clearly an effort to rationalise! Because of an intent to drive the probable point that the Quran has alluded to perhaps the precise number of planetary bodies in our solar system. A number which needs no human ratification in Astronomical sense of what is a planet and what is in the dwarf category.[60] If there is however a decision to disfavour any dwarf body from the present list being considered a planet in the planetary sense then the total count of planets equal to twelve as appropriate bodies to orbit the sun. In this case, herein attempted rationalising is promising. However, evidence in this verse to imply such rationalising is not compelling. The Quran need not propose an accurate figure either or it may also be setting its own standard for their numbering of how many planet bodies graduate as appropriate Planets. Wholesomely, they had to be seen eleven because Joseph had eleven brothers. Also, the interpretation of eleven stars for kawkab-s could be considered good enough; which allegorically appeared to him in a dream descriptive of the apparent reality surrounding him. This is a plausible interpretation for what a dreamer would dream being most likely of objects from prior gathered knowledge. Having said, considering Joseph's early learnings from his father Jacob via Abraham's remarkable enlightenment in astronomical sciences, it is not surprising that he knew what a planet was when he saw his famous dream still as a boy. Therefore, it is again plausible that eleven كوكبا (kawkab-ann) were truly planets in his knowledge figuring his eleven brothers, such that the crown in solar system, the Earth was honoured by its thirteen most prominent (eleven planets and the sun and moon) members. To rationalise isn't a sin; but idolising legends despite insufficiency of evidences in that direction amounts to a clear enormity. But still, I may be guilty of dogmatism, if I forced my variant interpretation in

---

[60] Mercury, Venus, Earth, Mars, Jupiter, Saturn, Uranus, and Neptune are counted as the major eight planets with addons from dwarf planets. Ceres is in the asteroid belt between Mars and Jupiter, while the remaining dwarf planets are in the outer Solar System and in order from the Sun are Pluto, Haumea, Makemake, and Eris.

this area where allegory as a dream is presented. Therefore, I will leave it to the readers' discretion to credit Joseph for his awareness of astronomical sciences taught by his father Jacob as a boy before betrayal by his brothers into Egyptian bondage. Unfortunately, insufficient Biblical accounts do not narrate complete storyline of Abraham's former years of prolonged efforts contemplating learnings before ascension into the praised records of the Lord. Times that preceded his Prophethood seem to be absent from the covers of the Bible. [61] But are enumerated in the covers of the Quran as discussed herein.

However, the Quran does clarify between a planet and a star to conclusively assert Abraham's gathering sure knowledge of these two astronomical bodies, even in contexts atypical to his story line. In verse 82:2 كواكب (kawakibu) which means "stars plus non-luminous bodies" is talked of as being subject to scatter at the unfolding of doomsday.[62] This makes full sense for all astronomical bodies, asteroids, comets, planets and stars to fall from their regular orbitals in unfolding of the celestial calamity as per Allah's command to collapse the first Cosmic sky. Whereas in verses 77:8 and 81:2 proper noun for stars is employed which is النجوم (an-nujum).[63,64] These verses are specific in talking of only "stars" becoming dim and losing their lustre. Therefore, it is seen clearly elucidated enough in the corpus of the Quran that كوكب (kawkab) is connoted to mean an astronomical body in whorl. The plural of which is not just the stars but also various non-luminous bodies in whorls. It is thus understood that stars having lost their brightness also scatter at the unfolding of doomsday. Abraham's comprehensions of the artistry in the creation of Almighty Lord Allah grew richer as he would interact with Angels in visits descending from celestial zones and thus his faith emerged

---

[61] It is a known fact that Kings in Christendom played a part in dismissing scrolls and manuscripts as a result Biblical collection is missing valid info that which the Quran guardingly has come to narrate. As the final revision from God, the Quran is seen to collect the missing bits and clarify the historical wrongs scripted in the covers of the Bible as an encompassing text over it. Quran 5:48 "And We have revealed to you (O Muhammad), the Book in truth, confirming that which preceded it of the Scripture (Bible) and as a criterion over it...."

[62] Quran 82:2 And when kawakibu (the stars plus non-luminous bodies) fall, scattering.

[63] Quran 77:8 So when an-nujum (the stars) are blurred (or dimmed, obliterated).

[64] Quran 81:2 And when an-nujum (the stars) turn quaggy losing lustre.

unfailing and even in the most intimate of tests, he stood elegantly committed (see Quran 37:100-108). This again praises his comprehension of the laws in the Universe and of the Earth which Allah had given him as mentioned in the very beginning of this section in verse 6:75.

For full story on Abraham, follow references from Quran given in footnote.[65]

---

[65] Abraham's full story, Quran references: 2:258; 21:51-72; 37:83-110; 6:74-84; 43:26-28; 11:69-76 and 26:69-85. Also, 3:65-68; 9:114; 2:124-140; 19:41-49; 22:26-31; 2:260; 29:24, 31; 51:24-37; 53:37; 14:35...

3.2 EGYPTIAN FEATS IN ASTRONOMY

Prophet Joseph's ascent in the Egyptian Dynastic kingdom promisingly explains ancient Egyptian heights in Astronomy. Not claiming that Egyptians weren't privy of such an emerging science prior to Joseph's own box of gifts. As a rising civilization, they were demonstrating applications of astronomical contexts to determine the spectacle rise of a few prominent stars such as Sirius whom they connected to the flooding of river Nile. At the beginning of the Dynastic period in ancient Egypt somewhere close to second quarter in the 3rd Millennium BC, Egyptians were already demonstrating a 365-day seasonal calendar. Also, the astronomically oriented Giza pyramid complex representation of Orion constellation, albeit a disputed theory is one among their plausible achievements.

Later, Ramesses II the Great Pharaoh of all times dated to have been born in 1303 BC is identified to be the one who encountered Prophet Moses (Arabic: Musa). A Hebrew from the slaves- fortunate to have been raised in Pharaoh's family after birthing amid slaves. This correlation drawn is important because of the immediate decline noted in their reign by their successive Pharaohs in Egypt after Ramesses II. This accords well with the Quranic accounts dealing with Egyptian historical facts. In popular sources, the decline of Egypt is dated to have commenced around 1640 BC.[66] Whichever of the period (date) is true, for the fall of Egypt's grandeur—perhaps Ramesses II is wrongly celebrated for his misjudged birth day. The Quran implies that this was due to Ramesses II's rejection of Allah's call to faith and due to his persecution of Israelites even after Moses – a Prophet, was sent to warn him of his transgressions. It was around this time that members of his chamber began to collect memoirs of another Prophet. Prophet Joseph who had served a tenure in their Dynastic history in the past.[67] Moses defended Israelites before this great Pharaoh of all times who besides

---

[66] Most information on Egyptian heritage in this section is sourced from Wikipedia. Wherever Biblical and Quranic perspectives are narrated a quick scriptural reference is given.

[67] Quran 40:34 And to you there came Joseph in times gone by, with Clear Signs, but you ceased not to doubt of the (Mission) for which he had come: At length, when he died, you said: 'No messenger will Allah send after him.' thus does Allah leave to stray such as transgress and live in doubt. (For full story see Quran 40:23-45).

rejecting faith, had declared himself as God, thus dropping the final straw for Almighty Allah to act. So, thereafter in a chase of Israelites Allah brought him, and his armies under the waves of the sea and drowned them.[68]

> Quran 10:92 "This day shall We (Allah) preserve you (Ramesses II) your body, that you may be a sign to those who come after you! But verily, many among mankind are heedless of Our Signs!"

The Quran implies to his body being preserved as a sign for latter generations to take heed from. Mummy of Ramesses II is now restored in Cairo's Museum. Great Pharaoh of all Ramesses II deserved a grand-scale funeral, which had been their tradition for honouring Kings; but, his moving from tomb to tomb after being scavenged from his drowned state for sparse veneration in the leftovers of Egypt, explains Allah's gripping him at his death because of his wrongly endeavoured chase of Israelites to subdue them back into slavery.

Tracing the dates, Joseph as a boy was traded for a miserable gain by his brothers because of his fondness with their father Jacob. Little Joseph had suffered at their hands many mocking and several physical injuries too; when they had pushed him into a well in the distant woods. Later they sold him like a slave for a petty price to a Caravan heading to Egypt. However, in Egypt Allah had a plan for him and He raised his status with the elites in the Pharaoh's kingdom. There he made several contributions by knowledge and wisdom taught by Allah. He was also very learned as a boy through teachings by his father from Abraham's heritage learnings. Per the Quran's story line, Egyptians then and the Pharaoh had become believers and shared faith in Joseph's Lord, after he had their backs secured from a drought that lasted for seven long years. Much fame came Egypt's way when many surrounding and distant lands were heralded at the news of Egypt's inexhaustible store supplies during the severe drought. Being a brilliant visionary and an administrative strategist, Joseph's management plan had thus made the Pharaoh's dynasty rise to prominence. Egypt's first civilisation transiting into this phase is considered as the Middle kingdom beginning 2040 BC at the end of 3rd Millennia BC. Considering Joseph's plausible adoption in Egypt

---

[68] Quran 40:45 So, Allah protected him (Moses) from the evils they plotted, and the people of Pharaoh were enveloped by the worst of punishment.

helping their transition from first civilisation, it makes complete sense that his niche contributions of knowledge had accelerated Egypt to accede to fore in the Middle kingdom as believers of the new age. Around about this time the settlement of his extended family in Egypt had happened. Then much later to Egypt's prosperous times the Hebrews had been living their lives toilsomely numbering for every thread of yarn. Being enslaved to the Pharaoh, fate had begun to bite them hard. Probably this reckoning of numbers was much later scripted as a prophecy of 400 years to Exodus in Genesis, as foretold in the scripture.[69] At least in my scholarship of scriptural studies Lord's spelling of 400 years or so for such a scenario is not scholastic of His narrative.

This number of precise counting of the years around 430 for that matter, synchronises the dates well enough from Joseph's adoption to the Exodus of Israelites (between 2040-1640 BC) and subsequently to mark the fall of Egyptian grandeur beginning 1640 BC. The high offices of Joseph, extended welcoming tokens to Israelites upon their first migration to Egypt and did not preclude them from positions of contributions. In these times before succumbing into long standing slavery, they had thrived in a long-privileged era of contributions favouring the grandeur of Egypt. At the time of their migration to Egypt, the original group of Israelites consisting of Jacob's family numbered just seventy people in census.[70] His aged parents and families of his eleven brothers made that humble number. In later years, their numbers had swelled and at the time of Exodus about 430 years or so they were amounting to 2.5 million despite their numerous male-born being killed in nurseries by Pharaoh. Of them were six hundred thousand strong men making up the infantry excluding women and

---

[69] Genesis 15:13 Then the Lord said to him, "Know for certain that for <u>400 years</u> your descendants will be strangers in a country not their own and that they will be enslaved and mistreated there." Also, it is counted with the Hebrews that in Exodus 12:40 the people of Israel had been living in Egypt for 430 years.
[70] Genesis 46:27.

children.[71,72] It therefore entails, it is more likely that Joseph had tarried with the Egyptians from times of their first civilization contributing to Egyptian heights giving them the needed leverage to flourish and to family in such massive numbers. Even if we conceded a much lesser and plausible census population of about a million and a half, perhaps empirical enough with women and children numbering significantly more, makes obvious sense for Pharaoh not to lose their cheap labour. Hence, he pursued the lucrative chase to be only overwhelmed by the avenging God.

The Egyptian Middle kingdom is dated to have ended 1640 BC after its 400-year mighty lead. Thereafter much later around 305 BC it was absorbed by Ancient Greek's Hellenistic kingdoms – The Ptolemaic kingdom. For a Millennium Egyptians had lost their historic trace after their once grandeur. During this early Millenia after Exodus Prophet David (Arabic name: Dawood) was inherited by Prophet Solomon (Arabic name: Sulaiman) who built a great kingdom in the holy land (Arabic: al-ardh-al-muqhaddasah).[73] After many generations of succession of Kings, this was later ended by a big blow by Nebuchadnezzar II in 589 BC.[74] This devastation of the grand house of Israel was for the sins of the Hebrews before the Lord of Moses, who had forewarned them not to break the covenant with Him after they had it ratified.[75] Historians who reckon Exodus and Solomon's kingdom a myth shoulder the burden to explain: who quelled

---

[71] When the Israelites left Egypt in the exodus, there were "about six hundred thousand men on foot, besides women and children" (Exodus 12:37). Also, see Exodus 1:7-10. Further, In the second year after the Israelites left Egypt, Moses took a census of the men in Israel able to fight—...The number of warriors was 603,550 (Numbers 1:45-46).

[72] According to undivided opinions of Biblical scholars such as this ref: bible.ca. The Hebrews numbered 2.5 million, an Exodus population.

[73] Quran 5:21 O my people (Jews), enter the Holy Land which Allah has assigned to you and do not turn back (from fighting in Allah's cause) and (thus) become losers.

[74] Quran 10:93 And We had certainty settled the Children of Israel in an agreeable settlement and provided them with good things. And they did not differ until (after) knowledge had come to them. Indeed, your Lord will judge between them on the Day of Resurrection concerning that over which they used to differ.

[75] Quran 17:5 So when the (time of) promise came for the first of them, We (Allah) sent against you (Jews) servants of Ours (Persian Nebuchadnezzar II) - those of great military might, and they probed (even) into your homes, and it was a promise fulfilled.

Ramesses II and his might in Egypt just about their heights of grandeur? And why did the Egyptian kingdom fall after all, around the corner parting Ramesses II the mighty Pharaoh of all... And in 589 BC why did the Persian Nebuchadnezzar II erode the grand holy temple built by Solomon from a distant march from Persia? If Israeli heritage wasn't worthy enough to inherit, they wouldn't have marched from Persia making mighty arrangements for distant warfare. In the absence of many gaping historical accounts of takeover of Pharaohs and Israelites, the slates are blank and Historians blind. Which therefore explains the historical disconnect with the rise of the new nation of Israel immediately after the Egyptian decline which (Solomon's legacy) had lasted for centuries preceding the emergent Greek grandeur between 5th-6th century BC and their classical era in 3rd century BC.

This analysis therefore credits Egyptian astronomy for having had a significant boost by Prophet Joseph's tenure in ancient Egypt when at a time they are documented for being adept at watching stars and observing the conjunctions, phases, and risings of the sun, moon and planets. With some rationalising in this section, a broader narrative is brought forward with evidences from the Quran verses and thus ambiguity is far removed. This then beckons claimers of reason to demonstrate their sincerity for the acceptance of it.

For full story on Joseph see Surah Yusuf (The chapter Joseph) numbered 12 in the Quran, and Quran 40:30-34.

Figure 7. *Venus reflected in the Pacific Ocean.*
Venus is always brighter than all other planets or stars as seen from Earth. The second brightest object in the image is Jupiter; Image credits: Brocken Inaglory, Wikimedia commons.

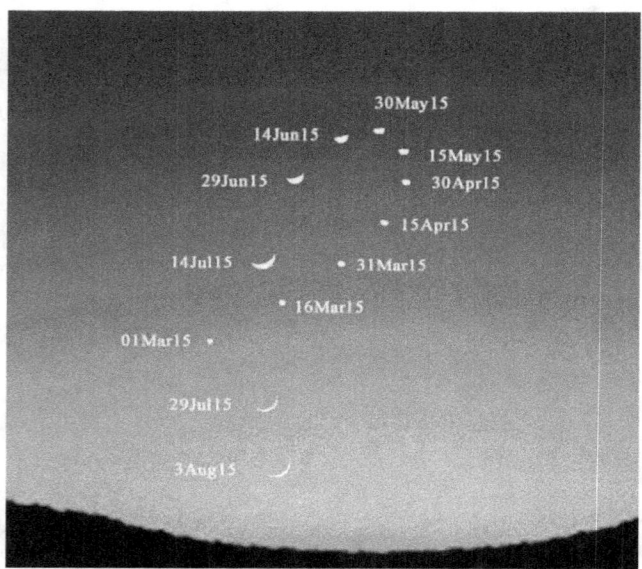

Figure 8. *Phases of Venus.*
Venus from March 1st to August 3 for 7:30 p.m. (local time). To show the phases of Venus on this scale, the angular diameters plotted are highly exaggerated. Image credits: Arvind Paranjpye, Nehru Planetarium of Nehru Centre Mumbai, India.

4 THE NEXT INTERFACE

Reliance on humans supposedly charismatic and intelligent or suspending decisions until they develop a majority consensus is not the correct premise to reason the truth of Allah's existence or even to analyse His message's worth. Undoubtedly you belong to Someone!

In the first two chapters we have seen in detail about the seven earthly skies and the make of solar system and its members. Now in the coming chapters it is time to journey beyond solar neighbourhood into the yonder regions of space. We start by defining the first major Cosmic sky sparsely mentioned in previous chapters.

> Quran 71:15-16 Don't they reflect how Allah has created seven skies overlapped. And has made the Moon amidst as lightsome. And had made the Sun an anchor.

These standalone verses deliver the full-basis to confirm the interpretation of a heliocentric system, not needing detailed sections of chapter 2 to cement this concept. As already seen in section 2.5 the earthly skies, prominently the sky of gravity (satellite sky) in their structural overlap house the moon. This is amidst the Magnetosphere and plasma tail of the Heterosphere on the dark-side of the sky where this area is found to stretch up to 1000 earth radii. Because of the Moon orbiting earth at about 60 earth radii, it finds itself located amidst earthly skies (except during its orbits of non-alignment) and from there it shines up as lightsome. Then, this whole system is referred to be anchored to the sun.

> Quran 37:6 Indeed, We (Allah) have adorned the **near world Cosmic sky** with an adornment of **bodies in whorls.**

> Quran 25:61 Blessed is He (Allah) who has placed in the sky starry constructions and placed therein (in one such starry grid) an anchor (Sun) and a lightsome Moon.

The verse 37:6 uses the Arabic phrase السماء الدنيا (as-samaa ad-dunya) traditionally translated bluntly in Islamic corpus as **nearest heaven.** Which simply means the sky of our world, in other words **near world Cosmic sky.** Because of the English word Heaven is synonymously used for Paradise, hereinafter an attempt is made in this work to rename it as the **first Cosmic sky.** Thus, in our apparent view from earth's reference it is a sky for our star i.e. the sun, our planet (earth and its skies), its satellite moon and this solar system cocooned amidst millions of other nearby star systems comprised in what is called as the first out of the seven major Cosmic skies. It is understood in astronomy as the interstellar sky for any given star cluster, like the solar neighbourhood for the sun. This sky however is but, only a respective sky. Several such first Cosmic skies makeup for different probable worlds. Thus, there are as many such first major skies for other potentially probable worlds including one for the planet of Paradise, a concept elucidated in good detail in this work. The first cosmic sky of our world is filled with building blocks of all sorts of astronomical bodies in whorls الكواكب (al-kawakib) and that of constellations of stars بروجا (burujann) meaning mansions of star assemblies in plural.[76] These بروجا (burujann) make up the many spectra of constellations firstly within the solar neighbourhood, and then in the expanse of the galaxy Milky Way which is the next Cosmic connect known as second Cosmic sky and then in beyond skies as discussed in section 6.1. However, in this sky, the sun is placed for earth's reference as وهاجا (wahhaja) meaning a glowing lamp and so is the moon as منيرا (muneera) which means lightsome. Further, as seen earlier in

---

[76] This is a unique word found in the Quran which is descriptive of star structures to any degree of detail or aggregation. The بروجا Burujann is the plural for Burj. Its meaning is easily explained like this: consider Baith which in Arabic means "a house." Now a building comprises of many houses. This is termed as برج Burj. Hence a Building comprising of several units of houses is termed as Burj. The megatall skyscraper Burj Khalifa in Dubai is descriptive of its Arabic name i.e. it is one building having multitude of houses. Now imagine mansions of buildings like multiples of Burj Khalifa. This is Burujann (a plural form) used here in the Quran.

this sky, the sun is made an anchor سراجا (siraja) meaning a loaded saddle such that our earth along with its satellite moon are gravitated to orbit around it along with the other known load of **astronomical bodies in whorls** (planets and other space objects) of our solar system. Here the sun cruises its path as is determined for it just like it is determined for other stars also placed in this sky as seen in section 2.4. Such is the exposition of earth's satellite sky interfacing with the first Cosmic sky as described in the glorious Quran.

For an understanding of boundaries of first Cosmic sky see **Figure 9**.

4.1 ASKING NATURALISTS

The just outcome of all known sciences and of rational thought has baselined a hard-fought yield i.e. to acknowledge the plausibility of the creator undisputed. Nothing of the evidence in the history of mankind did manage to remotely suggest otherwise. Therefore, "there isn't any burden to shoulder proof to disprove Allah's existence"; but rather it is the revulsion to belief in Allah which must be expediently explained. Often hateful distastes to belief in the Lord of the worlds result from the impulse of carelessness. Also, due to surmising self-sufficiency and the resulting pride. Commonly as it is today, revulsion to belief in Almighty Lord is stemming in the face of clerical misrepresentation of religion. Despite once inherited religious past, God has become delusional among latter subscribers. Even science researchers such as naturalists complexed by the mockery of clerics delve merely arranging ill-constructs of bad-logic and therefore imprudently join hands in framing rebellion to Allah.

> Quran 23:84-85 Say (O Muhammad), "To whom belongs the earth and whomsoever is in it, if you should know?." They will say, "To Allah." Say, "Then will you not remember?"

> 86-89. Say, "Who is Lord of the seven skies and Lord of the Great Throne?" They will say, "(They belong) to Allah." Say, "Then will you not fear Him?" Say, "In whose hand is the realm of all things- and He protects while none can defend against Him- if you should know?." They will say, "(All belongs) to Allah." Say, "Then how are you deluded?"

In plain rational sense Allah asks a rhetorical question: Who is the Lord of creation and that of the great throne? The response is given in favour of the Creator. He originates creation from nothing. In a verse, He asks a still closer question: *Quran 52:35 "Weren't they created from nothing or were they the Creators?"*

When you ask Atheists: "Where did we come from?" they merely answer, "we have come into being by a random chance." Elucidating this, naturalist scientists would narrate a deterministic model telling, from everything in the Cosmos to the realisation of life was inevitable: *"...our understanding into origins of life has come a long way, we know evolution happened by itself, ... we don't know how exactly it happened but you have a chemical evolution which creates these complicated molecules... and again I emphasize that we don't understand how it happened, ..., it becomes the first self-replicating molecule and once you get that then biological evolution is inevitable, so once you get life then you have evolution that's almost by definition of how life originates. …, life evolved probably about 3.6 billion years ago and there was no oxygen in the atmosphere when there were volcanic activity and lightnings when we probably had a lot of undersea activity and so to say that we can't make life today is completely ludicrous as a way of ruling out the origin of life 3.5 billion years ago."*[77]—Both scenarios are not further from being absurd: Firstly, the amusement drawn by using "randomness and a chance" is akin to deferment to enquiry in a scientific way, because, not only did we not come by chance, but there isn't a thing in the universe pointedly which resulted by a chance. For to credit- a chance, universe is so expansive and wide for not to have had anything coalesced at all. Physicists at CERN in Switzerland say: *"All of our observations find a complete symmetry between matter and antimatter, which is why the universe should not actually exist."*[78] *Scientists performing extremely precise measurements of properties: mass, electric charge and so on, have found no difference with matter and antimatter. At CERN's Antihydrogen Laser Physics Apparatus (ALPHA) experiment probed an atom of anti-hydrogen with light for the first time, again finding no difference when compared with an atom of hydrogen. Matter and antimatter by nature burst of pure energy leaving nothing behind- the most efficient reaction known to physics. Per Big Bang analysis equal amounts of matter and antimatter is produced – but that's a combustive mixture to annihilate itself for good, leaving nothing behind to make galaxies or planets or people: Quran 2:255 "… He (Allah) knows what is (presently) before them and what will be after them, and they encompass not a thing of His knowledge except for what He wills…."* Even if God granted us true knowledge concerning the subtle minutiae of matter and antimatter or how the universe has come to be with the seeming improbability, then in that case will God's existence come to fore to be questioned? This way of thought is ludicrous. Our knowing or ignorance in any field does not predicate validity to question God's existence. First, God is fundamental to consciousness. And second, His denial requires evidence, as in evaluating the principal to prove a creator less phenomenon. Antecedent creator is as simple as it could be—a fact.

---

[77] Words of Jerry Coyne, professor and an evolutionary biologist at the University of Chicago; author of 'Why Evolution is True' in Freedom from Religion - Interview published 20 May 2017 on YouTube: video time: 42-45 minutes.
[78] Title search: Universe shouldn't exist, CERN physicists conclude; News-Physics 23 Octo 2017 on cosmosmagazine.com.

Secondly, having claimed nature by itself to have aided origins of life; today as this discipline is in our firm grips, it is beckoning that we must demonstrate origins and creation of life in labs, it is indeed ludicrous not to commit for this endeavour. Even if the US congress won't public fund it, naturalists can afford to self-fund in parallel excellence to compete with the faithfuls in offering sacrifices & philanthropy, to enlighten the pupils of the land who have persevered in faith in their Lord of glory, despite the shabby impetus of trails casted to challenge their belief. This assertion if demonstrated with eye opening evidence by causing the flip of chemicals into biology will truly suffice. Thus far attained feats in biology have used the existing cell machinery to activate a synthetically generated virus. The cell is the key to life. So good luck making the key!

Just like diverging religious opinions (which is purely human, to dissent in any given field of knowledge) some popular minds researching modern science (including Lawrence Krauss) strongly suggest that the Cosmos has always existed, which accords with the truth of Allah's eternal existence. As told, Allah originates newer origins in the pocket grids of the universe from scratch.[79] But the premise of an antecedent cause to Allah has been rendered as farfetched by new science revelations commenting that such an instant alluding to the existence of a precursor cause in the historic distant past never existed.[80] Naturalist sided deterministic models inherently fault when postulating initial stable conditions (preceding Big Bang), which can't explain origins. They at the same time advocate that there was never a time when there was absolute initial-nothingness (at time=zero). This way they fumble to concede that there has always been something. Freewill is at the source to result determinisms. Determinism—is a philosophical theory that holds that all events are inevitable consequences of antecedent sufficient causes; often understood as denying the possibility of freewill. But without freewill constructs won't result therefore deterministic order and forms can't be explained. Freewill is encompassing of knowledge to result in the new makings and not determinism. In fact, determinism is a blocker to knowledge. It is clearly Allah who is free of all wants and who is not questioned for anything but enforces determinisms and justices indicative of *Quran 21:23*

---

[79] Quran 35:1 All praise is to Allah Originator of the skies and the earth...
[80] Title search: 'No Big Bang? Quantum equation predicts universe has no beginning' on phys.org.

"*He (Allah) is not questioned about what He does, but they will be questioned.*" Thus, reliance on humans supposedly charismatic and intelligent or suspending decisions until they develop a majority consensus is not the correct premise to reason the truth of Allah's existence or even to analyse His message's worth. Allah is prevalent in deeper consciousness and His revelations such as the Quran is as clear as day. Let me strike an example, suppose I enact in best manners to imitate a familiar voice to you, despite my outstanding deliverances your unique enquiries to me will prompt me to succumb from imitation revealing my true nature. Similarly, the Quran will clarify to you in a convincing way that it is not Muhammad or someone else but—Allah Himself Who resonates to you. It is perhaps ones' cognisance that will truly aid to identify demi-gods living or dead and to discern and deselect among the cheerful science-researchers from those that are pre-emptive of truth. In the worst case, one may prolong the wait anticipating expert opinions and finally lose the opportunity for belief.

4.2 NEAR AND GREATER SKIES

The only verse in the Quran that spells and distinguishes Cosmic skies as seven from its earthly seven skies is: *Quran 65:12 "It is Allah who has created seven (Cosmic) skies and from the Earth similar number. (His) command descends through between them so you may know that Allah is over all things competent and that Allah has encompassed all things in knowledge"* from surah 'taulaqh' a chapter named -divorce or partition-. The technical inference is after admonishing proper ways to part between spouses for valid reasons having laid down it's divine guidelines, Allah then reveals in verse 12, information regarding partition between the cosmic skies from earthly skies in His artistry. In the Quran, there are nine places where Allah mentions seven firmaments and only one of them i.e. 65:12 refers to cosmic skies and the other eight refer to earthly skies. Therefore, those eight quotes in the Quran where the number of skies is stated as seven, needs to be studied with respect to the earthly skies which the Quran uses to elucidate pertinent information or to allude to its wondrous artistry in which we are warmly cocooned. Encompassing words in scripture are context specific. So are words such as "every pair" in the scripture in verse 11:40 concerning Noah's times. Noah of the ark, whose flood was a localised phenomenon unlike the widely misunderstood global deluge for the phrase "every pair of creatures" which obviously did not include ocean dwellers nor hoppers of Australia neither the many hundreds of thousands of creatures of the greater earth. It was aimed at drowning disbelievers who rejected his call which is a prophecy commissioned by God for every people resident of a region and not an act to submerge or punish the whole of earth. This story relates to Noah's people, the disbelievers of that age who had lived in lands of a scalable geography in a region of the earth from where God commissioned Noah to gather all animal species belonging to that ecosystem to rehabilitate the region post deluge. After Noah's flood, Allah led humanity to spread on earth and gave them scriptures. This was the transition from ghetto framework of early humans into various nations dispersed on earth: *Quran 36:41 "And a sign for them is that We carried their forefathers in a laden ship."*

God gave scripture to Noah and followed it by revelations to succeeding many generations up until Muhammad in simple yet elegant prose in many vernacular languages to those who were becoming various nations of the world.

> Quran 11:48 It was said, "O Noah, disembark in security from Us and blessings upon you and upon nations (descending) from those (believers who boarded) with you. But to nations We will grant enjoyment; then (if rebellion stages) there will touch them from Us a painful punishment."

Not just my opinion but several scholars researching comparative studies in religion conclude: Noah of the Bible and the Quran is referred as Manu among the Vedic religions of India. Where a scripture is called Manu smriti—known as the law book, still extant although shredded by adulterations: *Quran 42:13 "He (Allah) has ordained for you of religion what He enjoined upon Noah and that which We have revealed to you (O Muhammad), and what We enjoined upon Abraham and Moses and Jesus- to establish the religion and not be divided therein. Difficult for those who prefer desires is that to which you invite them. Allah chooses for Himself whom He wills and guides to Himself whoever turns back (to Him penitent)."*

Thus, the Quran's emphasis on creation narrative is driven to induce reflection in the immediate surroundings prompting us today and as it lay over its history of manifestation to motivate scores of meditating men, inspiring many and irking many, since its emergence preceding middle ages from 610 A.D. Encouraging men to ponder upon the wonders of Allah not from far but from within and in very near horizon around us. Therefore, in the Quran much attention concerning earthly skies is drawn for a closer retrospection. And then the Cosmic skies are perhaps briefly elucidated as found in verse 20:4 *"A revelation from Him who created the earth and **greater Cosmic skies**."* The Arabic word العلى (al-oulaa) in this verse refers to the major Cosmic skies. Verse 65:12 conceivably is the only verse that is found in specific to number the major Cosmic skies as being seven and then telling that the earthly skies are similarly numbered. Having said, sufficient details have been given concerning major Cosmic skies as discussed in chapter 6, except nondisclosures deemed as the matters of the unseen which the Creator hasn't unveiled.

4.3   SEVEN EARTHLY-SKIES ALIKE IN NUMBER TO COSMIC-SKIES

"The sky" in Quranic Arabic is denoted by the word السماء (as-samaa) which is a singular noun form. For several skies it uses its plural form السماوات (as-samawat). But in case of earth, the Quran has always used a singular noun-form only—الأرض (al-ardh) meaning the earth. In the entire Quranic corpus, a plural for the earth is unfounded. An Arabic word for <u>earths</u> does not exist in Quran although it does in modern evolution of the Arabic language. Since Vedic times, religions have exhibited evidence of succumbing to fallacies and misinterpretations due to not corroborating information based on investigations and inquiry. Hindu Vedic interpretation of 'Patala' is one such derived concept resulting from the mythical misconception of the Vedic scripture. 'Patala' on Wikipedia will tell you how much humans could imagine in those ancient times. Folklores concocted around the idea of earth having a similar number to that of skies has since broadcasted many fantastic stories of creatures living and suffering on six other layers below the earth's surface. Thus, subsequently the scripture has been wrongly taught to reinforce suggestions made by earlier fabulous interpreters narrating folklore blinded by whimsical fellowships of fancying superficial knowledge.

In Islamic traditions, there are many alleged Hadiths made popular amongst Muslim scholars to promote the concepts like Hindu folklore of 'Patala' alleging that broad-scale life on all seven 'earths' to exist. Perhaps arranged deeper below the earth's surface layer that we live on. Unfortunately, despite eye-opening scientific discoveries blind followers aren't persuaded to reform their stale understandings or even to eclipse superstition by reflecting upon their primary scripture afresh, despite many present excellences like of Vedic subscribers. Amongst blindly led Muslims, hearsays, traditions and culture, and ancient fabrications overpower reflections of their principal scripture—The Nobel Quran. Therefore, it is a hard battle to imply value to Allah's own narrative. Here is the evaluated translation, importantly, of the most misjudged verse of the Quran in this area of knowledge.

> Quran 65:12 It is Allah who has created seven (Cosmic) skies and **from the Earth** similar number. (His) command descends through between them so you may know that Allah is over all things competent and that Allah has encompassed all things in knowledge.

In Arabic syntax of the verse above, the conjunction من (min) is abnormally interpreted by traditionalists, even by my many respected commentators due to lack of required reflection required in this area. Normally من (min) means 'from' which results as its first translation. And it does not translate as 'of' or 'about' to begin with, unless it is context driven. If the Quran were to employ عن (un) instead of من (min) then the conjunction in that case would translate as 'of' i.e. 'of' the earth or 'about' the earth. Then there would remain no further scope for any newer arguments to buy-in anymore negotiations such as me criticising its traditional mistranslation and interpretation. Two obvious reasons qualify the case concerning this verse: firstly, a plural form of the noun 'earth' is not found in the Quranic corpus and secondly, use of a conjunction من (min) in this verse, meaning 'from' (i.e. 'from' the earth), dictates that this needs revised interpretation. [81] The argument that above re-evaluated translation for this verse is the valid translation, is founded on this very premise. Nevertheless, there is a mountain of evidence from within the Quran as elucidated in this work to support this interpretation of mine and I have on impeccable tokens tried to clarify this in chapter 1 and as will be seen in chapter 6.

Thus, the interpretation follows that Allah has created seven Cosmic firmaments, which are seven major skies as found explained in the Quran. Then Allah has created "from the earth" a similar number of earthly skies. The command of Allah descends through between them and effects earthly beings. Such that we may know that Allah is over all things competent and that He has encompassed all things in His knowledge.

---

[81] Tracing reading errors in ancient scriptures: Semitic languages, where Semitic is a derived name from Shem a son of Noah (source: Biblical). Various nations of Semites which descended making many cultural groups after the deluge had derived languages on common threads. Modern analysis of the structure and construction of languages such as Arabic, Aramaic, Hebrew and Sanskrit endorses evidence of a common derived pattern typical of Semitic languages. God's communication to humans is deciphered or interlaced on grammar and syntax found common to these languages. This analysis therefore shows absence of enquiry has resulted in fabulous interpretations of God's scriptures since Vedic times in strikingly similar patterns.

The following verse further explains the creation of planet earth and its skies as a separate act of creation in the grid locations of the major Cosmic skies:

> Quran 2:29 It is He (Allah) Who has created for you all things that are on earth as complete systems; also, His design comprehended the skies, for He perfected them into seven firmaments; and of all things He has perfect knowledge.

As seen in verse 25:61[82] that our sun is a star in the interstellar sky and planet earth was made to anchor to this light emanating burning saddle. Therefore, per verse 2:29 cited above: The earth and its blanket seven skies were originated, *Quran 35:1 "All praise is (due) to Allah- Originator of the skies and the earth…"*, in a grid in this patch of sky anchored to the sun situated in the first Cosmic sky in Orion Arm in Milky Way. Hence, the understanding that earthly skies are made alike in number to that of the major Cosmic skies that Allah has constructed does arise as the only credible interpretation of the verse 65:12 discussed herein.

---

[82] Quran 25:61 Blessed is He (Allah) who has placed in the sky starry constructions and placed therein (in one such starry grid) an anchor (Sun) and a lightsome Moon.

4.4 EXTRA-TERRESTRIAL PLANET EARTH AND SKIES

In the farther outdoors of celestial space, there is another planet of this exact nature. Still a bigger one as told in the Quran.[83] One that is keenly—the Paradise earth. Popularly known as the Heaven where the promise of eternal life stands aspiringly awaited. Exoplanets thus have been occupying human imagination ever since Adam, the first of our kind, to whom in the launch of our race was promised eternal-life in the Paradise home after his brief faithful period here, in the Orion home.

> Quran 11:106-108 As for those who become wretched, they will be in the Fire. For them therein is (violent) sighing and wailing. They will dwell therein for a time that the skies and the earth will endure, except as your Lord wills for your Lord is the (sure) accomplisher of what He plans.
>
> And those who are blessed shall be in the Garden. They will dwell therein for time that the skies and the earth endure, except as your Lord wills a gift without break.

In the above set of verses the Quran is clearly indicative of a separate planet and extra-terrestrial life where we will be led for afterlife. Disbelievers in the Lord of universe, Idol worshipers, hypocrites plus bad performers or sinners amongst believers too will be flung into a pit of fire in that planet. Unlike others those who truly had even an atom's weight of faith in God alone, conceivably one day stand a chance to be released from it after meeting penalties which will atone and purify them. This very large fiery pit of hell is perhaps a geological crust plate, with fearsome, desolate and isolated scorching activity of ponds of volcanoes guarded by nineteen Angels: *Quran 74:27-30 "And what will explain to you what Hell-Fire is? Naught does it permit to endure, and naught does it leave alone. Darkening and changing the colour of man. Over it are Nineteen (Angels)."* They, stern Angels will ensure punishments are meted to its dwellers and none manages to escape. The believing, good performers will reside in Paradise off and away from the infamous pit of suffering. They shall be admitted into the gardens wherein rivers flow, away

---

[83] Quran 57:21 Be ye foremost (to seek) Forgiveness of your Lord, and a Garden (of Bliss), the width whereof is as the width of skies and earth, prepared for those who believe in Allah and His messengers; that is the Grace of Allah, which He bestows on whom he pleases. And, Allah is the Lord of Grace abounding.

from hearing any of its sighs and wailing: *Quran 21:101-102 "Those for whom the good (record) from Us has proceeded, will be removed far therefrom. Not the slightest sound will they hear of Hell; what their souls desired, in that will they dwell"*; except on a request to dialogue: *Quran 7:50 "And the companions of the Fire will call to the companions of Paradise, 'Pour upon us some water or from whatever Allah has provided you.' They will say, Indeed, Allah has forbidden them both upon disbelievers."* This constructs a clear idea of a planetary earthly body with its respective seven skies, in the regions of celestial space. The verses further indicate that both inmates will abide therein for durations until that planet and its skies endure for; or as Allah wills it to exist for eternity. The latter is the case as we learn from the Quran.

Scientists say, many exoplanets are worse, either too hot or too cold… just one earth-like planet is missing from our telescopes which they are hopeful of finding. At the University of Arizona in Steward Observatory Mirror Lab, engineers are working on the next gen super telescopes with the largest mirror size of 8.4 meters across and combined with 7 such segments needed for the ground based Giant Magellan Telescope (GMT) scheduled to begin in 2022. The plan is to use this to directly image earth like planets with 10 times sharper images than the Hubble telescope in the quest to search for stars containing an earth like planetary kin. The astronomers who have dedicated their quest believe it's a matter of time before we discover truly earth like planets. We are commonly discovering planets they say, but not quite like the earth.[84] But sending space probes to almost every planet and to some of their moons in the near reaches of the solar system isn't buying us any good learnings of discovering Allah's mercy. Soviet's robotic probe with heavy armours is trashed by the extreme atmosphere in Venus so lovely from earth up-close this goddess of love is hideous, its global warming we now know has gone wild, it is a model of sisterhood depicting hell. Of the many hells that surround us and of the many dangers that loom in the celestial space from where the Voyager in interstellar space is drawing neat sketches for us to be rather submitting to Almighty than evicting faith. [85] Atheistic foundations have

---

[84] Traveling to Other Galaxies - The Search for Earth like Alien Planets - Space Documentary; published 23 Feb 2017 on YouTube: video time: 41-43 minutes.
[85] Journey through the universe beyond the speed of light- Space Documentary; published 9 Feb 2013 on YouTube.

answered their own missions to win successes neglecting the promised earth by Almighty Lord. The Paradise planet, perhaps you may locate it on GMT in near future yet you won't be able to flee to it even if it be well within the realms of Milky Way: Quran *29:22 "Not on earth nor in sky will you be able (flee) to frustrate (Allah), nor have you, besides Allah, any protector or helper."* The fact is Almighty Lord, Allah is telling us that despite our missions surged with determined will to evict the imminent solar scorching, it is not possible to evade His final justice.

4.5 COSMIC SIMILARITIES

Other than natural phenomenon itself, the Quran is a sure reminder of Almighty Allah and a source alluding towards His otherwise improbable wonders in creation. Allah has enhanced human inquiry by blessing us with superior faculties. Thus, it is expected of us to amass learnings through keen observation of ourselves and of the surroundings. At length, when Allah would deem necessary to setup the promised resurrection day for us on earth housed in this solar neighbourhood; He will scale-down Cosmic spans to transport us to the best determined exoplanet where the concept of afterlife is awaited to commence:

> Quran 50:31 And Paradise will be distanced near to the righteous, not far.

Therefore, our earth with its skies as discussed thus far is not a unique world. But it may be so in the solar neighbourhood respective of the first Cosmic sky within an arm of our spiral galaxy. Milky Way has tellingly thousands of such interstellar local star clusters. They are known as first Cosmic skies respectively or the first major skies of respective worlds. If Allah so deems, it may become plausible in future to find life in them citing if worlds and life of whatever kind exist in them. Such first Cosmic skies may be many as understood as the habitable exoplanets where life is perhaps found in other types or kinds or as will be ordained by Allah in futuristic eons to come. Every galaxy inherently is a second major Cosmic sky made from collections of many such starry constructions or local clusters, identified by various arms or spurs tucked within them. The earthly skies with the respective first Cosmic skies are thus encompassed. It ensues for the Paradise earth that the chances are it is perhaps located in not very distant galaxy, but in a near interstellar sky within Milky Way. Inferences to this concept are further discussed in section 6.9 and 7.6.

> Quran 3:133 And hasten to forgiveness from your Lord and a garden as wide as the skies and earth, prepared for the righteous.

The planet where Paradise of eternity is created is in fact much larger than our planet earth here in the Orion Arm. Quran 57:21 is also found to repeat this information. It is alluded that Paradise earth is as wide as the width of our planetary earth up to its skies in altitude. By an approximate measurement, the distance of our earth's sky up to its natural satellite -the moon- calculates the planet of Paradise by scaling it to at least three-times as large as Jupiter. Thrice the size of Jupiter is clearly massively large. It's a huge planet of Paradise, for the righteous to inherit apart from cutting out room for a continental sized pit to host sufferers, known as Hell. The planet of eternity is said to sport similar varieties of flora and fauna as found on planet earth with exceptional brands oncourse included. It is said, the inmates of eternal garden, while conversing hint to have eaten fruits in their earlier life to which the Quran tells, they will be given in similarity.

> Quran 2:25 And give good tidings to those who believe and do righteous deeds that they will have gardens (in Paradise) beneath which rivers flow. Whenever they are provided with a provision of fruit therefrom, they will say, "This is what we were provided with before." And it is given to them in similar likeness. And they will have therein purified spouses, and they will abide therein eternally.

Quality in Paradise is of premium and par excellence. Among the speciality are the *'hoor'* damsel creation for us to intimately pleasure-with, created as a variant female human species of Paradise earth, with big lustrous eyes, intact on heels, brocaded, adorned, and splendid in body tone.[86,87]

> Quran 56:35-38 Indeed, We (Allah) have produced the women of Paradise as a new species. And made them virgins. Dedicated: equal aged for the companions of the right.

We will be made to spouse with them on exclusivity and they will remain bashful and dedicated: *Quran 37:48 "And with them will be women limiting (their) glances, with large, (beautiful) eyes."* No more pangs of promiscuity will there be in that ambience to abuse our dignified personal space.

> Quran 76:19-20 There will circulate among them young boys made outstanding. When you see them, you would think them (as beautiful as) scattered pearls. And when you look there (in Paradise), you will see pleasure and a great dominion.

---

[86] Quran 44:54 Thus. And We will marry them to fair women with large, (beautiful) eyes. See also, Quran references 56:22 and 52:20.
[87] Quran 78:33 And full-breasted splendid companions.

And another variant for speciality are the *'wildaan'* young boys for services of hospitality and for hotel management. As told, the inmates of Paradise are free from prohibited lusts. These asexual boys are to offer royal services despite their sight-overwhelming lure, they will only magnify the realm of dominance for the inmates of Paradise. Cited from a couple of surahs from Quran 76:5-18 and 56:17-26 here is a prose- As to the Righteous, they shall drink of a Cup (of Wine) mixed with Kafur. A fountain making it flow in unstinted abundance. Reclining (in the Garden) on raised thrones, there they will be in just the right weather conditions. And the shades (of the Garden) will come low over them, and the bunches (of fruit) there will hang low in humility. And amongst them will be passed round vessels of silver and goblets of crystal. They will determine the measure thereof (according to their wishes). And they will be given to drink there of a Cup (of Wine) mixed with Zanjabil. A fountain there, called Salsabil. Round about them will (serve) youths of perpetual (freshness). With goblets, (shining) beakers, and cups (filled) out of clear-flowing fountains. No after-ache will they receive therefrom (like our fermented bottle-o do to us), nor will they suffer intoxication. And with fruits, any that they may select. And the flesh of fowls, any that they may desire. And (there will be) Companions with beautiful, big, and lustrous eyes. Like unto Pearls well-guarded. A Reward for the deeds of their past (life). Not frivolity will they hear therein, nor any taint of ill. Only the saying, "Peace! Peace!"[88]

And about what our Women will get in Paradise is answered like this *Quran 32:17 "No soul knows what is hidden for them of the things that are to give delight for the eyes as rewards for what they have performed."* Both Men and Women will be pleased with our possessions in Paradise and it is a promise on Allah. And Allah does not fail in His promise.

---

[88] Description found in Quran 76:5-18 and 56:17-26.

5 FIRST-COSMIC SKY DRESSED IN STARS

Beauty has many enigmas that startle worthy beholders; when a
tricked maiden exhibits herself, she is merely stolen for fashion and
not for any lasting prize.

Consider the following verse:

> Quran 25:61 Blessed! is He (Allah) who has placed in the sky
> **starry constructions** and placed therein an anchor (sun) and a
> lightsome moon.

In the above verse, the Arabic word بروجا (burujann) is a plural. Traditionally this has been translated as "**constellations.**" But it is far more than just constellations.[89] It accurately reads as mansions of starry-formations or array of starry constructed mansions. It best describes the varied spectra of constellations in the expanse of celestial space. However, like the word السماء (as-sama) meaning the sky, بروجا (burujann) also has contextual connotations for what it could mean and how it may translate. In case of السماء (as-sama) we have seen it can be invoked to mean any among the seven earthly or the seven Cosmic skies.

---

[89] Line diagram (**Figure 12**) shows configuration of Astronomers' consolidated constellations. It drafts constellations divided in sectors across the breadth of Milky Way and beyond with the locus of the earth as the observation centre. Milky Way is classified as part of Sagittarius constellation, Andromeda a near distant major galaxy is consolidated as part of the Andromeda constellation in the modern adopted methods of constellation classification.

Similarly, بروجا (burujann) could connote to characterize any of the following: Firstly, the etymology of traditionally drafted sectored classification of the night sky as constellations but within the realms of solar neighbourhood, see **Figure 12**. Secondly, the arms of a galaxy with its mansions of starry-structures. Thirdly, many units of galaxies in a galaxy cluster known as the local galactic group containing the likes of Milky Way, Andromeda and fifty plus more galaxies in a cluster. Fourthly, a cluster of many local galaxy groups grouped into mega structures called Superclusters like Virgo Supercluster which are among the largest known structures. Fifthly, the likes of millions of Local Superclusters forming the observable filaments of the universe. All these connote بروجا (burujann) according to the Quran.

In verse 25:61 cited above the word بروجا (burujann) is connotatively used for the second case to point towards Milky Way's sub-galactic elemental starry constructions. In each arm of Milky Way there are many different mansions of starry-constructions. This inference is drawn due to information pointing that sun is placed amidst one such starry constructed site that is rich in star-density. High-density starry establishment is typical of our solar neighbourhood indicative of verses in Quran 37:6 and 67:5 which tell that *"Allah has certainly beautified the **sky of this world** with glowing lamps (bright stars) and space objects in whorls"*, see **Figure 9** and **Figure 10** to grasp what our immediate Cosmic sky contains. It is in a way unique and teaming with most prominent stars in our galaxy.

Boldly launched and endeavoured modern astronomical discoveries are found to unambiguously complement this Quranic information. GAIA Sky is a project developed in the framework of ESA's Gaia mission to chart one billion stars of our Galaxy. It is a real time 3D astronomy visualisation software. You can look for firsthand information by browsing blogs on galaxymap.org. Having perceived what experts have colluded from thus far intense observations, it is obvious that the Quranic information communicates accurate descriptions of the greater Cosmic reality. It is clear from images shown via GAIA software that the sun is placed in a sky local to it packed with high density and rich stars which are otherwise not visible to observers on naked eyes.

Excerpts from GAIA blog citations: We know already that stars are not distributed randomly but are found in vast concentrations, much like flocks of birds or schools of fish on earth. These concentrations range in scale from systems with multiple stars, clusters, and associations all the way up to spiral arms. There is structure at every scale. The TGAS data set released as part of Gaia DR1 in September 2016 contains positions and parallaxes for 2057050 stars, about 85% of the 2432906 stars in the Tycho-2 catalogue. This is the first time that parallaxes (and hence distance estimates) have been available for such many stars. We get this effect because Gaia can see more stars that are closer to the satellite and fewer stars further away. The TGAS data is known to be incomplete. A more complete and accurate data set extending far beyond the solar neighbourhood into the spiral arms and the galactic nucleus is scheduled to be released in April 2018. For now, TGAS is the best data we have for mapping the Milky Way. TGAS contains parallax data for about 2 million stars, which is most of the Tycho-2 catalogue.

5.1 NAVIGATION BY STARS

> Quran 6:97 And it is He who placed for you the stars that you
> may be guided by them through the darknesses of the land and
> sea. We have detailed the signs for a people who know.

Naturalists in their criticism of natural phenomenon are often caught off-guard by these verses. Like some naturalists who concerning anatomy call the prostate gland—a design by a moron engineer, what better way would a designer resort to, to give weaknesses which is a part of His grand plan? Naturalists failingly argue concerning the design of unimaginable to scale distances and sizes of stars and galaxies built in celestial skies. The fact emerges more prominent upon observing that it is not just humans, but admittingly migratory birds and animals are guided by stars in their respective journeys through their continuous days of night-journeying. Like the Polar bear who at the switch of seasons swims to the Artic through continuous nights guided by stars!

> Quran 16:16 And by (natural) mechanisms. And by the stars
> they are (also) guided.

Mankind despite long history of once traversing caravans through wildernesses; sailed seas and even Airplanes guided by stars seem to now have conveniently forgotten purpose in their design. Due to the burden of theological context implying Allah's hand in the Cosmic existences. Naturalists tend to gain ground by arguing over its clarity of purpose by inferring to point failure in their design for any connected need. Also, in the wake of disfamiliarity owing to modern innovations of gyroscopes and auto-pilots, naturalists fail to fathom purpose behind simple and elegant designs in the sky. If the scales and distances weren't as they are, its purpose would have at least demonstrably failed to effect wisdom-based responses from animal and bird-kinds.

Necessarily the first Cosmic sky couldn't be more shrunk in scale to serve for the true need of night-navigation for the many creatures on earth. It ensues for the size of this interstellar sky its galactic centre is again precisely calculated which comes to explain the size of the Milky Way itself. It is said that our solar system neighbourhood is situated in the goldilocks-zone within our galaxy. The best probable place to be in a galaxy for life to be possible is for the earth and the Sun to be in a range where they are now- within the goldilocks-zone. Thus, one can imagine the investments made in design in the construction of the Milky Way.

This proportional invocation of material existence within our galaxy is perhaps an easy to hop edge for the human and Jinnkind to see beyond. Enabled by a far simple invention of a telescope we humans have begun gazing many of many superclusters of galaxies. Self-centrism in religious philosophy is blameable for colluding the idea of us being at the centre of this Cosmic realisation. Is this as vast an existence a need for humankind? Certainly not. In fact, we weren't at all at the centre of this bold archaic design. Long before humans, Angels have rejoiced in this wide artistry praising Lord by His hymns. This is the unique prowess of Almighty that His creation is made to exalt Him. Angel-kind in their stately physical sizes of far greater scales travel through Cosmic pathways over long distances, as detailed in section 7.5. Such beings dwell in this universe's expansive realm and work for what their Lord commands them to. Well clearly planet earth is not at the centre of the observable universe and neither were we conceived to become central to it; but it so appears due to our measurements of the distant-light reaching to us from all directions giving an impression that we're squarely placed at the observable centre. However, we are making a definite parable for what must come in future in Lord's grand plans. But as told in: *Quran 21:17 "Had We (Allah) so intended to take a pastime, We could have taken it from (what is near) with Us- if (indeed) We were to do so."* This verse dumps us as in a far-off county from what is considered positionally near to Allah. Here though the reach of Allah is not alluded to diminish. Allah is ever near, regardless of His positional farness; this science is further elucidated in chapter 7. Therefore, it was required with Allah, that we not be placed in proximity to Him and hop an edge peeping into His abode such that the test on recipients of freewill be proceeded with, to determine who among us believed in Him unseen of His territory. Thus, the frontiers of His residence are not made observable to us, that's how far we are casted. Given these distances, Angels cruise stretched Cosmic spans to reach us and ascend back to God over thousands of years of celestial travel in speeds above human grappling keeping things really hidden.

As Angels are directed to us in this immense Cosmic artistry, for their navigational needs our interstellar solar neighbourhood is geared for homing them. An array of pulsars that are identified by their unique timing of electromagnetic pulses help identify our solar position both in space and in time by potentially extra-terrestrial intelligences (Wikipedia). Because pulsars emit very regular pulses of radio waves, its radio transmissions are not susceptible to corrections. Such is the design of God that from similar things in this case all-stars, He brings about stark differences for aimed purposes. Thus, Stars in Orion spur host destinations for Angels launched into descent from Cosmic skies from beyond. We cannot fully know in certain terms the many secrets God has brought about within the first Cosmic sky to work out our affairs; we may only conjecture as we ascend slowly learning from our past misunderstandings. God is only known by His attributes as enshrined in the scripture. So, we should know Him by codecs enshrined in scripture instead of lensing our ideas of philosophy upon the divine. Clearly, we haven't taken the right course to find the god-particle in our stately laboratories.

Excerpts from adsabs.harvard.edu: Within Milky Way are found, an estimated seventy thousand potentially observable pulsars with luminosities in the range 0.3-10 mJy-kpc2. They note, the distribution in galactocentric radius rises smoothly through solar location towards the galactic centre and peaks inside 6 kpc (kiloparsec), again in contrast to the observed sample which peaks near the sun; in an article titled- The galactic population of pulsars.[90]

---

[90] 1 Parsec = 3.26 light years (ly). Light year is the distance covered by light in one year in vacuum travelling at a speed of 299,792 km per second.

5.2 INFORMATION NETWORK

Among the popular geographical spots in the world for star gazing is South Australia. It bears the impetus to surprise the naked human eye. Living here has clearly been my best star gazing experience, not far from city limits. It shows you the sky filled with the beauty of stars akin to wearing a telescope gadget. But it is not just the astounding beauty that these stars are constructed for. The stars constructed in the interstellar sky serve many other purposes. One of them is as said in this verse:

> Quran 37:6-10 Indeed, We (Allah) have adorned the sky of this world with an adornment of bodies in whorls—And as protection against every rebellious Satan. (So) they may not listen to the **higher assembly** and are pelted from every side— Repelled; and for them is a constant punishment. Except one who snatches (some words) by theft, but they are pursued by a meteor, piercing (in brightness).

The earth has never stood abandoned by Almighty Allah as naturalists wilfully assert. Rather, it is served-to in immaculate ways which we do not fully comprehend. The unseen and undetectable community of Angels who work orders from Allah on the earth and in its skies, are a designated staff for this world to bring about matters concerning human affairs in the realm of our reality. Angels encompassing our first Cosmic sky exchange information and communicate orders coming from Allah channelled via above Cosmic skies. The الملإ الأعلى (al-mala-il-āla) translated as **higher assembly** in context of these verses is the voleries of Angels relaying scripts from the above Cosmic skies stationed in each major firmament to the assembly seated in Orion, our first Cosmic sky adorned with the dense network of stars.

> Quran 67:5 And We (Allah) have certainly beautified the **first Cosmic sky** with glowing lamps (bright stars) and We have made them as deterrents for Satan-kind and have prepared for them the punishment of the Blaze.

According to the Quran, like humans, the non-human receptors of freewill are the jinn (in plasma dimension) who via their science advances, try to intercept information beamed to Angels staffed within the first Cosmic sky.[91] This is true for the rebels from among them while those aware and obedient jinns use science for just ends. Angel-kind as information bearers and off-loaders of orders from Allah by their elusive makeup habituate regardless of star magnitude and perhaps even at potentially light gravitating black hole locations.[92] Therefore, Heliocentric-habitability for earthbound information transmission is realised as the main-post for Angels on earthly routine activities. Therefore, stars constitute travel corners or sign-posts and communication filter stations for Angels. It is hard to know truth from far! Thus, revelations from God in fact make the absolute truth as Allah has also said that He is immanent and further communicates to us in our own languages. Verses from Quran 53:7-10 from surah the star (an-najam) allude to this fact and is discussed at length in section 7.6. From this area of interest perhaps Astrology at its delusional peak with previous civilisations had slipped into human cultures by the dearly sought jinn-friendships for rebelliousness.

The information from Almighty Lord meets its human receptors in three ways. Allah's ways to communicate with different things and beings differ by their underlying relation with the Cosmic disposition.

With us however God communicates as follows:

> Quran 42:51 And it is not for any human being that Allah should speak to him except by inspiration or from behind a partition or that He sends a messenger to reveal (scripts, tidings and warnings) by His permission, what He wills. Indeed, He is Most High and Wise.

The first type وحي (wahi) i.e. inspiration of non-scriptural type is that which is planted upon the consciousness of a person and is directed as a precise idea rather

---

[91] Quran 6:112 Likewise did We (Allah) make for every Prophet enemy: Satans, among men and jinns; inspiring each other with pleasant seeming discourses of deception...

[92] Title search: Researcher shows that black holes do not exist' on phys.org. Regardless of the known truth of black holes and gravity bending light; Quran has beyond ambiguity depicted positions of sites to annihilate stars, indicated of verse 56:75 discussed ahead. Call it a Quasar or the Blazar or a Super massive black hole feeding spree.

than verbatim scripts and the mind resolves the conceived idea further, which then generates factual memory of it as it is brought into action. Also, this mode of inspiration reaches many righteous people too (benefitting mankind) other than Prophets, for example Moses's mother was inspired to many actions concerning her child: *Quran 28:7 "…suckle (your child), but when you have fears about him, cast him into the river."* This sort of communication does not involve verbatim deliverances, it is a concept tucked into a mind via bio-plasma interface from an elusive dimension from God Himself, wherein time-dilation is not a detrimental factor, more detail on process of time-dilation involving وحي (wahi) or يوحي (yu-hi same as yu-ha) is discussed in section 7.5.

In the second type, to Moses who was uniquely honoured among humanity, Allah spoke direct by a partition. When he perceived a bush burning up the mount, Moses closed in to gather a fire brand but was astounded... he had heard Allah's audibles fall on his ears direct.[93] This way of reaching out is very subtle. The energy enshrouding the bush coming from the Lord appeared to burn it, but rather the bush and the fire played a transistor. Turning incoming energy from God into audibles to Moses: Quran *4:164 "…and Allah spoke to Moses with (direct) speech."*

In the third type وحي (wahi) in the sense of revelations by a carrier Angel known as an Angel Messenger comes upon orders from God, who acting upon His directions, channels scriptures via bio-radio communications. The bio-radio communications of all listed methods adopted are perhaps the strongest and heaviest to bear upon our anatomy; communication engineers researching the narrow bands of radio communications perceptible to living organisms such as the unnerving of animals during Earth Quake scenarios and neuro-specialists who could further help decipher this interface ability by analyses of forthcoming plausible states of neuron-configurations in our brains could usher more light in this area of science. This investigation in identifying bio-suitable radio signals could be the helping first steps in understanding revelations to Prophets of God.

Unlike the second handled case of direct-speech which is via an interactive channel (burning bush: transistor) from the elusive dimension in which God enabled real-time interactions where Moses was connected by God Almighty and

---

[93] For glimpses of this story refer to Quran 20:9-36; 28:29-35.

where the chance for time-dilation stood eliminated. The first & third types makeup a one-way information imparting channel and there is again difference in extended usage between the two. The first use for وحي (wahi) is for non-scriptural inspiration as seen earlier upon the conscious mind of a person, which is by bio-plasma interface caused by God Himself to men and to several other creatures (floral, faunal and Jinnkind). And in elusiveness also to Angels. On the other hand, وحي (wahi) by the scriptural premise is a revelation by a carrier Angel endued in wisdom to impart teachings and this is very strenuous. Herein the target is ingrained with fuller words of a script message via bio-radio communications. In this the revelation is by نزّل (nazzala) meaning sending down of scriptural words. Verbatim words then are neatly positioned in the memory regions of the target as unique etchings which the Prophet is then able to retrieve with ease, as is the case realised with all human Messengers, receiving verbatim words from God known as scriptures. These do not make mere constructs requiring memory generation surrounding the idea as in earlier case of وحي (wahi) of non-scriptural inspiration, but the words are engraved and is recallable to be announced before masses in recitals or simple sentences given in same languages conversed by the target people. This mode of communication is affected by dilation. This is because even instant direction (communication by first type) by God to Gabriel to reveal scripture requires preparation of the script by Gabriel for circumstances unfolding in human domain which must be readied (extracted from an excerpt from the mother book) or packaged for delivery in clear decipherable words.[94,95] This is precisely the reason why you'll find in the Quranic narrative surrounding Prophet's movements having to deal with various situations among people, verses were given reminding of the past tense with prefixes of resumption particles such as 'And when', 'When' and imperative verbs such as 'Say' etc. And as these admonishments had to be remembrances of the historic happenings with Prophet Muhammad for generations of believers to come and thus it had to be a script-based revelation rather than constructs or

---

[94] Quran 43:3-4 Indeed, We have made it an Arabic Quran that you might understand. And indeed, it is, in the Mother of the Book with Us, exalted and full of wisdom.

[95] Quran 56:77-79 Indeed, it is a noble Quran. In a Register (excerpt) well protected. None touch it except the purified (Angels).

idea of inspiration given to navigate life-situations via bio-plasma channel from an elusive dimension.

Jinnis due to their experience in mastering bio-plasma channel for their rebel some interface with humans to ooze whispers, they have come to study and build techniques to intercept the radio-communications by Angels of God. Scripts of instructions are neat algorithms. These divine distillations have to be sent in codecs via radio communications which dictate the future of worldly affairs as long-time dilation is involved for its coming from God's own assembly involving several years of space travel. When the Angels in the exteriors of first Cosmic sky recept these codecs having to deal with revelation or algorithms for general duty work orders, they secure its modulation. And then channel it via their sophisticated network to deliver it to Gabriel. Due to densely packed starry network buzzing incoming communication, Angels seated inside this grizzly cocoon in close locus of solar locations do not get to recept it; thus, Angels organize sophisticated networks for delivery into interiors after intercepting communications from the exterior boundaries of this antenna (see **Figure 9** and **Figure 10**). God does communicate to Angels via وحي (wahi) inspiration but for elaborate duties and for scriptures Radio communication is acceded as a clear channel. Thus, starry network in this patch of the sky is compacted with high density stars. The leaked straits however by this dense starry network are susceptible to reception by the acutely trained jinnis. Therefore, communication amidst Angels over wide area distribution within the realm of first Cosmic sky is again susceptible to prying if it is done via means where jinnis are learned to tap. This is where the artistry of starry network with its numerous space objects kicks-in to play a role. (**Figure 9** and **Figure 10**) not only show us the mind-blowing architecture but easily informs us of the buzz the stars are able to generate and their potential to perform signal-jamming of the Cosmic communications directed towards the first Cosmic sky. This is an area where many fierce attempts by jinnis to listen to some of this are resourcefully prevented. And as and when they do manage to get some bits, they are in que to surmount chases via space objects of flame casted by Angels: *Quran 15:18 "Except one who steals a hearing and is pursued by a clear blaze."*

Prior to emergence of human civilisation, Angels located in this part of celestial sky were summoned by Allah for a new creation when Adam was created. Iblis (later nicknamed Satan) then a devotee of Allah rivalled our position and did not buy in Allah's plans for Man. Since then he has managed to deceive and train many of his kind to launch contortions over human minds in almost every area of interest than just to show counterevidence to realising existence of God Almighty. But as always, Allah extends a firm hand hold by revealing scriptures to those sincere in reasoning. Telling that truth demands: justice be done to falsifiers, as His Prophets don't play pretence but are tied to communicate what God deems for us to know. Since the beginning of mankind's free-willed activity on earth, voleries of Angels under command of Gabriel have descended to near locations of solar system clearing rebel some jinn stations to facilitate (undisclosed) timely reach of God's pristine light of guidance and His favours & retributions alike as they were needed to justify many determinisms. [96] Muhammad in seventh century was chosen and he was taught to declare: *Quran 38:65 "Tell (them)- I'm undoubtedly made a Warner..."* from the Lord of Cosmos and the earth and all between this, telling of the Quran as a great news. And further telling, "that I have no knowledge of the proceedings with the Angels in divine discourses with God, it is but a revelation that I have received that I'm made a clear Warner."

For glimpses of this story, follow references from the Quran given in Surah Suad (the chapter named after the syllable Suad) reference verses: 38:65-88.

---

[96] Determinism: Also a decree of God to interfere into human affairs at the level of wider population. It mostly concludes in death & destruction and can also result in suffering and losses.

5.3   ANGELS AT SIRIUS

> Quran 86:1-4 By the sky and the **night comer**. And what can
> make you know what is the **night comer**? It is the **piercing star**.
> There is no soul but that it has over it- guard.

The Arabic word الطارق (at-tauriqh) meaning the night-vigil connoted for
a knocker during night, alludes to the one who watches. It comes from the root
word طرق (tauraqha) meaning knocking, pounding or beating. The Sirius star
system or simply Sirius is a widely known star since ancient times. Being closer
to earth it is also the brightest Star with an apparent visual magnitude of minus
1.46. The apparent visual magnitude of a star is a measure of its brightness as
seen by an observer on earth. It is expressed as a number and the smaller the
number, the brighter the star. For Sirius it is -1.46 the lowest for any star in our
reference of observation. The next in line is the Canopus (313 ly away) with a
value of -0.72 followed by Alpha Centauri (4.4 ly away) with -.27 and so on. Sirius
is revered in many human cultures. The Egyptians connected it to the flooding
of river Nile. And the later to come Greeks around one Millenia BC, connected
Sirius with harsh summer days. The Polynesian dwellers of the southern
hemisphere had connected it with the approaching winter. Such is Sirius, a
prominent and mesmerising star in the night sky. Sirius aligns away from
appearance in the night sky approximately for seventy days and then dawns
again as the one on night-vigil. Shining at from 8.6 light-years' (ly) distance; a
light year equalling nearly 6 trillion miles is in fact the nearest star easily visible
to the unaided eye.[97] Classified by astronomers as an A-type star, its surface burns
at about 17,000 degrees Fahrenheit(F).[98] With slightly more than twice the mass
of the sun and about slightly less than twice its diameter; Sirius still puts out 26
times as much energy. It is considered a normal (main sequence) star, meaning
that it produces most of its energy by converting hydrogen into helium through
nuclear fusion.

From May to sometime after midsummer Sirius and the sun are in conjunction so that
the sun's greater light blocks the visibility of Sirius. The heliacal rising of Sirius occurs

---

[97] Light year (ly) is the distance covered by light in one year in vacuum travelling at
a speed of 299,792 km per second.

[98] The sun is about 10,000 degrees F.

when the star and sun are sufficiently separated. So that—for the first time in seventy days—Sirius can be seen on the horizon just before dawn. In the northern hemisphere, this occurs in mid-to-late summer in the mid-northern latitudes such as most of the U.S. Sirius rises in the southeast, arcs across the southern sky, and sets in the southwest. In December, you'll find Sirius rising in mid-evening. By mid-April, Sirius is setting in the southwest in mid-evening.

It is now known that what the naked eye perceives the star Sirius is actually a two-star system– the dominating Sirius A and a small, faint companion star known as Sirius B. This little companion also known as 'The Pup' was an earlier mighty main sequence star, even bigger than Sirius A, but now is a dead star called a white dwarf, an earth-sized ember too faint to be seen without a telescope (**Figure 11**).[99] The two stars are gravitationally locked onto each other and follow one another galloping in an elliptical orbit drawing a far end and a shorter end to an ellipse. The distance separating them is on an average just about the distance between Sun and Uranus. Due to its fierce vibration sequences and probable pulses in closer proximity than Saturn is to Sun at 8.2 astronomical units (AU), Sirius A exhibits pounding effect on the smaller star which the Quran aptly describes this behaviour by the root word "**to knock.**"

Quran 86:1 By the sky and the **Knocker**.

This effect is more prominent when the two-approach closer on the ellipse wherein the massively sized Sirius A exerts forces which nearly tear the pup from its orbital as it advances towards it. Their closer approach also generates huge magnetic storms between them and both stars begin to spin faster because of tidal forces getting stronger. Regarding the once massive Sirius B it is postulated here that before it became a white dwarf, it underwent an implosion ejecting incinerating energy in the near to earth celestial space owing to its closeness. A detailed discussion on Sirius elucidating its historic dynamism as interpreted from the Quran is covered in section 7.6.

---

[99] A white dwarf is what stars like the Sun become after they have exhausted their nuclear fuel. Near the end of its nuclear burning stage, this type of star expels most of its outer material, creating a planetary nebula. Only the hot core of the star remains.

However, it can be argued that apart from Sirius, another type of star called a pulsar best describes the knocking attribute given to Sirius. A pulsar is a highly dense, magnetized, rotating neutron star or white dwarf (Sirius B is a white dwarf), that emits electromagnetic radiation at rates of one thousand pulses per second. Pulsars are known for their small sizes with extreme densities and continuous radio beats of pulsation. But the argument is in favour of Sirius for many reasons and while Sirius B's pulses aren't typical of Pulsars. Firstly, from syntax and context, the proper noun identified in Arabic as marifah (a known object) is the substantive communicated in these verses by using the proper noun النجم (an-najam) meaning the (known) star, made known for its attributes. Secondly, the active participle الثاقب (as-saqhib) meaning piercing, penetrating and boring is typical of Sirius to allude to its proximate to us super brightness boring a channel straight down to earth to subdue evil of rebellion in its path en route to earth. It is clearly pointed at a known object in the night sky- the famous Sirius. The fact that the verse mentions 'a' star referring to its attributes as a knocker and piercing bright in the sky disqualifies pulsars which are notoriously far and detectable only through radio waves.

> Quran 86:1 By the sky and the **Knocker**.
>
> 2. And what can make you know what is the **Knocker**?
>
> 3. It is the **piercing star**.
>
> 4. There is no soul (human born on earth) but that it has over it- **guard**.

Now that it is established that Sirius is being referred here, it is noteworthy to ask, why so? The answer is in the 2nd and 4th verse. Sirius is alluded to knock and clear the rebel jinns efforts who pry Angelic communication en route to solar post. It is a knocker from where affairs are drawn as it is at Sirius from where Gabriel is in command leading his armies of Angels to work-out human affairs per divine instructions. It is immediately indicated in the verse where Allah says, "every human soul born" has over it- guard. The word **guard** besides meaning protector in cases preventing from fatalities per Allah's decree also means recorder of all actions and words, in command under their captain seated at Sirius.

The period of Gabriel's arrival to Sirius and crowds of arrayed Angels gathering at its horizon is a close subject to modern astronomy because of the dynamics involved with Sirius B's implosion. The implosion is described in the Quran in verse 53:9 by the verb قاب (qhaba) which also means to empty its fill. In modern astronomy, it is understood that on a norm red-giant phase of a star's life lasts for really longer periods, millions of years, and clearly not for a few tens or hundreds of years. But given a quick guzzle of matter by a hidden dimension like say creating a swift vacuum, the star is exhausted of its fuel rapidly showing burning red only for a brief time, akin to in binary star systems sucking the more massive star out of its fuel. Or a Fast-Evolving Luminous Transient (FELT) type supernova, which only lasts for a few days. FELT is a rare supernova phenomenon that we have come to realise lately in our never-ending learnings of what possibilities of events may exist in space. Any of this type or stuff we yet do not know is plausible in case of Sirius B by Gabriel's take on it from an elusive dimension erupting incineration targeting en route towards solar system. This scenario could possibly breathe life into the colour controversy surrounding Sirius showing red between 3rd century BC to 150 A.D. by poets and astronomers alike except for Chinees who described it white (Wikipedia). However, there are a couple scripture based probable scenarios to date this historic event:

1. First scenario, Gabriel descended at Sirius at the fore of Adam's assuming life to procreate and labour in the earth (after the expulsion from the garden). Which indicated effecting of the test for humankind. And since then Gabriel has seated this position revealing scriptures to all Prophets. Therefore, Sirius B turning into a white dwarf is thus dated.

2. In second scenario, it could possibly be around the dated colour controversy, when Christ (Arabic: Maseeh) was gearing up for his mission. For it is said of Christ that even in the Gospel holy spirit (Gabriel) was by his side: *Quran 2:87 "And We did certainly give Moses the Torah and followed up after him with messengers. And We gave Jesus, the son of Mary, clear proofs and supported him with the <u>Holy Spirit</u>. But is it (not) that every time a messenger came to you, (O Children of Israel), with what your souls did not desire, you were arrogant? And a party (of messengers) you denied and another party you killed."*[100]

---

[100] Also, Quran 2:253, 4:171.

In section 7.6 where this subject matter is discussed in detail its historic happening is predicated with the first scenario discussed above not making the colour controversy discussed in Wikipedia derail this narrative.

It is read that Jesus (Arabic: Isa) prophesised in the Gospel of John 14:26-31 of the Advocate and Prince of this world as coming and the Father will have him teach men all that Jesus had stood for. Then to Muhammad, Gabriel showed his stately physical form, introducing the Lord's kingdom. Referring to this the Quran tells in 42:52 that before this you- O' Muhammad, knew not what was revelation and what was faith, but Allah made it a light and guidance that you may lead men to the straight way.

> Quran 42:52 And thus We have instructed The Spirit (Gabriel) toward you (O' Muhammad) by Our command. You did not know what is the Book or (what is) faith, but We have made it a light by which We guide whom We will of Our servants. And indeed, you do guide (others) to a straight path.

Trust is another fundamental law operating in the universe. Evolution biologists for their poor comprehension of modern physics must invest their trust with the community of physicists in their systems, procedures and for their numerous assumptions. Same is valid for palaeontologists or for that matter any other discipline. Trust clearly mounts up huge stakes! The best criteria to invest trust in is for demonstrated honesty; by such as those who conquer biases and peer set notions and free themselves from the snares of vested interests like backlash from mainstream circles. Muhammad in his life amidst a small populace was well known for his uprightness, they called him Al-Ameen (the trust-worthy). Until when he declared that Allah had assigned him to warn against blasphemy, they utterly disbelieved in him. Which divided his society, and this part of history from then is well documented. Almighty Lord signatured this assignment with the show of Gabriel to Muhammad, a claim which he can openly make amidst people to summon them for faith unto Allah Almighty. There couldn't be a better opening for a people who were cut-off from prophecy and faith. Then again trust was needed to believe. Starting with this, people gathered gradually and then on a mountain of reasoning was garnered to the faith of Islam to last until judgement day.

5.4 STAR FACTS ANALYSED

Sirius B was once in its more massive size with about 5 times the mass of sun before its transformation into a white dwarf. Popular science sources establish Sirius B to have become a white dwarf 124 million years (MY) ago after shown a red giant for 25 MY, which is challenged here. The following section explains the approaches in calculating the age of stars and their accuracies. This is required here to entertain the possibility for the age of Sirius's implosion to be revised based on this information. It is also important to understand this to differentiate the established facts from "forced on facts" from among popular science data. Observation of a star spectrum, luminosity and motion of a star are some indicators to determine the age of a star by comparing to standard models. Undoubtedly the standard model for comparison is our sun. Its size, luminosity, colour and spin serve as a pre-calibrated clock for reference. Radiometric dating such as uranium-lead and potassium-argon decay are common isotopes used to sample meteors. This in turn helps determine the age of our sun, as meteors are considered remnants of solar creation. The gaps in sampling are tantamount to assumptions, small errors get quickly magnified for larger timescales. Sampling of archaeological specimens (of smaller timescales) has resulted in considerable errors by best-practiced methods. Isotope of Carbon (C) decaying into Nitrogen (N) C-14/N-14 for best results works with materials less than 50,000 years old and this has accuracy issues. It is important that students of science and commons be educated of its inconsistencies in (field) measurements instead of simply concreting the view that dating is absolute and perfect as in a lab. It is simply not the case for various specimens dated thus far. The Aboriginal remains in lake Mango in Australia and famous Himalayan cave samples shows wide scatter in data plotted from tests carried out by different agencies (Wikipedia). Which strikingly explains what amounts of error can accumulate in dating much older archaeological specimens. Dating meteors is even more complex due to many unknowns involved owing to their calamitous process of formation (discussed in next section).

A new technique called Gyrochronology involving measuring the rate of a star-spin is said to show dependency on star's age along with its variation in colour. It thus surfaces as a good indicator to help determine age of a star within an uncertainty of 15%. Also, as the luminosity increases for stars growing older due to increase in diameter, which again no doubt is a good indicator of a phase shift in the life of a star. Luminosity change via interferometry is how Sirius is aged to 225 to 250 MY.[101] Therefore, by calibrating 4.6 billion years (BY) to the birth year of our sun via meteor sampling, stars are compared to our sun for their mass, luminosity, colour and spin to back-track star facts. Acutely cruel but, an essential straw to seek light... concluding that more massive stars have shorter life span, observations in a way become preventative of fact finding. Thus, stars equalling sun's characteristics are favourably bestowed long lives and more massive giants are not so favoured to long life. Alluding that Sirius B now a white dwarf since 124 MY after shown a red giant for 25 MY isn't an outcome of a resulted mystery act, but a well understood phenomenon. However, the veracity of this claim after best efforts to deduce numbers around star facts by methods of observation drawn from data gathered over the previous century explains much insufficiency among the substantiated information collated in modern Astronomy. This clearly is impending future observational ratification as this data set perhaps is amounting to enormity. These are by far a few notoriously used techniques to determine ages of stars. The baseline notion of more massive stars in a generic sense colluding faster to death is critiqued herein. A quick comparison of star facts[102] from prominent stars botches the generalisations used in Astronomy to crunch numbers concerning ages of stars in a backward substitution method as follows:

1. See (**Figure 13**) for Betelgeuse, Rigel and Bellatrix. All three are from the Orion constellation which may have birthed at the same time being in the same clutch located at 642.5, 772.6 and 250 ly away from the earth respectively. Betelgeuse at 11.6 solar mass is said to be aged for only 8-8.5 MY and is now a red giant. While Rigel at 17 solar masses is still a BV type super giant with more or less same age as Betelgeuse (8 MY).[103] As seen in this example, there is no linearity between a star's mass and its

---

[101] The life and times of Sirius B, by Ken Croswell | Published: July 27, 2005 on astronomy.com.

[102] For star facts: browse astronomytrek.com.

[103] Internet search result: A B-type main-sequence star (B V) is a main-sequence (hydrogen-burning) star of spectral type B and luminosity class V. These stars are from 2 to 16 times the mass of the Sun and surface temperatures between 10,000 and 30,000 K. B-type stars are extremely luminous and blue. O type stars are the brightest and among the young stars.

age. If there was any linearity, going by popular opinions the more massive Rigel must have long set into advanced red giant phase and should have had a much shorter life span which is not the case.

2. Bellatrix at 8.4 solar mass is aged for 20 MY despite the low side difference in mass it must at least have progressed to becoming a- A, F or G type star if not a red giant at 20 MY of age, but it has not. It is still a B-type giant star (see **Figure 14**). Rigel & Bellatrix are evidence enough that there is great insufficiency in back-tracing star ages despite birthing in the same clutch. A standard model like sun cannot be taken and applied to all just like age of one human being cannot be taken as a standard to figure out life spans of humanity. This clutch of Orion (**Figure 13**) is conclusive in telling that stars are like men, some die young and others in old age. There isn't a generalised life cycle that is Astronomical worthy an equation for estimating the star life cycles.[104]

3. Further, Hadar (Beta Centauri) distanced from earth at 390 light years in Centaurus constellation is a triple star system, meaning they are born from the same cloud as triplets and are aged at 14.1 MY and have all-stars in B-type classification. Two are blue giants at roughly 10 solar masses and one is a blue dwarf at 4.6 solar mass. The two massive stars of the triplet resembling closely to Betelgeuse in solar masses, again must have grown old to their red giant stages already. Clearly having similar solar masses has not resulted them into having the same age!

4. Another analogy from within the Orion constellation is Mintaka a bright giant B type main sequence star, this star is in fact a complex four-star system (with three stars orbiting around Mintaka). It forms part of the Hunter's Belt along with the stars Alnilam and Alnitak (**Figure 13**). At 24, 30-64.5 and 33 solar masses respectively; allegedly aged for 14.1, 5.7 and 6.4 MY and at distances of 1200, 2000 and 1260 ly are easy to see in the night sky. Mintaka and Alnitak are the brightest young (O-type) stars in

---

[104] Not just my ardent opinion. But, for expert opinions see scientificamerican.com. Title search: "How do scientists determine the ages of stars? Is the technique really accurate enough to use it to verify the age of the universe?" Responses by, Stephen A. Naftilan, professor of physics in the Joint Science Department of the Claremont Colleges and Peter B. Stetson, senior research officer at the Dominion Astrophysical Observatory in Victoria, British Columbia.

the entire night sky. The star Mintaka at 24 solar mass (over twice the mass of Betelgeuse) and aged at 14.1 MY announces that Betelgeuse is dying too soon. Once again, the mass of Star has failed to keep linearity with its age. And, therefore the Hertzsprung-Russel diagram (**Figure 14**) showing lifetime indexes on stars can be concluded to have tremendous error factors.

5. Canopus aged 30 MY, a dubious number though (**Figure 14**). It has not been studied well therefore the difficulty in determining its age. Albeit at 9.8 solar mass distanced 313 light years in Carina constellation, it is the second brightest after Sirius in the night sky. Canopus has a B-V colour index, is a blue-white, indicating it is essentially white, although it has been described as yellow-white, again dubious. A white bright giant of spectral type-F dubious again, tracing Sirius closely in brightness index at -0.72 from a great distance away upsets the standards set of star facts of Betelgeuse and of Sirius. Canopus lives on whereas Betelgeuse is found dying. And if placed besides our sun, Canopus would blaze with brilliance of 14,000 suns[105] and Sirius to sun would just do at 20 times over. With such brilliance in brightness and mass trailing closer to Betelgeuse it is nowhere close to showing a red giant.

6. Most star facts listed here of Rigel, Bellatrix, Mintaka, Alnitak from Orion; then Beta Centauri from Centaurus and Canopus from Carina weep for Betelgeuse! In this astronomical predicament how can we be sure that stars depicted in **Figure 14** have been accurately charted for lifetime indexes? Clearly our persistence will amount to arrogance and pride!

Why is Betelgeuse dying while elder and more massive comrades in Orion constellation are spewing adrenaline? It's graduation to dying is a determination of Allah. It is plausible that a Cosmic pathway traverses en route to Sirius from Orion, for Orion constellation is where it is inferred that Allah had His throne parked while fashioning planet earth, a concept further elucidated in section 6.9. The line of stars leading to Orion are pitched in to avert jinnis from priceless

---

[105] Because of Canopus's great distance from earth and Sirius's proximity to earth, Canopus takes a position next to Sirius in the brightness index in the night sky as seen from earth. It is like a headlight a good distance away or airport beacon miles away.

listening grids. Bellatrix is closest to earth at appx 250 ly and Betelgeuse is closest to the Orion base at appx 643 ly from earth. Therefore, stars in Orion constellation occupy an important position in barricading rebel jinnis operations from affording an access to 'الملإ الأعلى (al-mala-il-āla)' the high assembly. Allah infuses life in stars as He wills and others, He races them to death to accomplish purposes of protecting His interests by graduating them to barricade the illegal activity in areas of restrictions. Having no need to draw more comparisons, it is very clear that there are grave inadequacies in Cosmic observations. The techniques are strewn with many loose assumptions like making our disproportionate sun as a benchmark star. With predicaments such as those discussed now, it is evident that Astronomers are battling with mighty tasks great distances away. Further, sciences of radiometric dating established for sampling meteor debris to determine age of sun brings about an inflation in assumptions inherent within such techniques. It is very likely that these seed enough constants to yield exaggerated figures like to start-with number crunching of sun's age as a benchmark to figure out ages of other distant stars. This we have seen already, is nothing short of a disaster. Hypothesis and facts are quite different things, they are like in religion hearsays and true revelations from God Almighty. Uncritiqued—scholarly interpretations and perhaps much thoroughly corroborated interpretations such as the one presented here to supplement and explain Allah's narrative yet falls short of truth as human endeavours are de facto fallible. First of all, in the scale of truth Allah's revealed verbatim words beyond ambiguity make the truth and as these foundation statements tell us of facts, no enhancement of its meaning endeavoured by us is without the wide impetus for furthering research in the quest of truth but maintaining fallibility. Lord of Sirius, Almighty Allah alone knows the star facts of Sirius's birth and its earlier full-companion star's time of collapse into a white dwarf.

5.5 DATING

Dating is an acclaimed, robust technique to date many samples in several contexts. Half-life of isotopes is among all other techniques, the best worked versatile technique to date substances as understood with a margin for error. However, the important limitation is real time (field) measurements which test our stand on being humble enough in our claims. In scrutiny of this technique scientific consensus favours potassium (K) decay into argon (ar) as more reliable for measurement than isotopes of carbon & nitrogen because of the ambiguousness involved in the calibrated (pre-set) value of C-14 in the atmosphere owing to a number of factors from Cosmic radiation to carbon cycle changing proportions of C-14 in the environment and also by influences due to industrialisation. With 5730 ± 40 years for its half-life its limits to age samples range from 500 to 50,000 years.[106] Giving it a window of 10 times factored and multiplied for an effective dating range of substrates. Some researchers put the upper limit to 20,000 years after which they say: carbon technique does not work well.[107] Therefore, in determining age of earth and sun isotopes with higher half-lives are needed such as K-40/ar-40 and ar-39/ar-40. However, the field again is much wider and more mysterious!

Prior to the formation of solar nebula this zone was a gas cloud. Perhaps not all gas but dust and accreting clay, probably from the remains of formerly exploded stars or from the gas and dust blown afar by galactic winds. These materials before coalescing into products of our solar system had an age of their own. This is among the detrimental factors in pinning sun's age. Argon is an inert gas which does not chemically react with other elements. And thus, as it forms by radioactivity in a freshly cooled rock it gets locked in its interiors. However, it is based entirely on the porosity of the rock type. Sample substances from this rock when measured tapping ar-gas in vacuum and weighing for left over K gives the estimate of age of the rock. Most rocks are porous which means there are definite pathways for new argon forming to leak. Also, when subject to heat the rocks

---

[106] Applying Carbon-14 Dating to Recent Human Remains, NIJ Journal No. 269, March 2012 by Philip Bulman with Danielle McLeod-Henning; https://nij.gov/journals/269/pages/carbon-dating.aspx.
[107] National Centre for Science Education (ncse.com) guided by its scientific advisory council and a board of directors promoting science.

soften expanding the pores and giving away for argon-leaks, which amounts to inappropriate estimation of ages. If the rocks are severely heated or melted, then the radioactive clock is reset as all of argon-gas is risked leaving the sample. In my opinion from porosity index, measuring granite rocks (known as least porous rocks) on earth from mountain formations is a good way of dating earth, but otherwise even closely contesting other igneous and several metamorphic rock types are not suited using this method. With samples picked from moon discussed under "lunar-granites" a concept criticised by other scientists as an attempt to vagrancy,[108] like to infer the improbability of locating granites outside earth or at least in the moon or in the asteroid belt or from meteors, the chances of collecting a suitable sample for K-ar dating is grim and fading. Moreover, the burnt meteor samples hitting ground or samples collected from surface of asteroid (itokawa) by Japanese hayabusa mission does not make required worthy samples owing to porosity and other unknowns. The best dating of earth is possible by working to improve accuracy of sampling granites found on earth which will give the age for early earth formation. But dating sun is another ballgame due to missing substrates and the issues discussed thus far, where evidence is shown to mount for a gross misappropriation of its age.

The half-life of K-40 disintegration is dated to 1.25 billion years. In determining the half-life of K-40 several parameters are worked to fix process limitations. Further in this attempt numbers (by math and empirical data) are assumed and corrections forced. Such as the linear logarithmic correction for zero thickness substrate layer for avoiding consuming of disintegrating electrons by neighbouring particles, where every decimal counts in magnifying or reducing half-life estimation; then the fudge factor included to account for stray counts of electrons due to not collecting in the detector; in a lab experiment by students a factor of 20% is applied to reach a value of acclaimed half-life of K-40; accounting for 8 missing events from 10 shots giving a 500% boost in the counts per minute of disintegrations.[109] If these corrections are kept to minimum values then the half-life of K-40 could be drastically scaled up from one billion to 10 billion years

---

[108] Title: Graphic Granite from the Moon by Authors: Warren, P. H., Taylor, G. J., Keil, K., & Kallemeyn, G. W. Journal: LUNAR AND PLANETARY SCIENCE XIII, P. 839-840; adsabs.harvard.edu.

[109] Radioactive Half-life of Potassium-40; prepared by Paul C. Smithson, Berea College, based on Postma et al.,2004.

and other way round if it's increased slightly it could be reduced from one billion to 100 million years which amounts to considerable deviation in the window for measuring ages with K/ar as the selected isotopes. Both these scenarios then toll the legitimacy of dating substrates in arriving at a date of 4.5 billion years to formation of earth.

Thus, besides repeatability of field measurements, there is very poor reliability on present calibrations owing to many assumed parameters to fix reading errors; even laboratory numbers could not be the absolute-truth. Techniques practised in lab are at times exclusive to workable samples. And degenerative field measurement is quite a different area for arriving at a correct estimate for the age of earth. Which again has limitations in our understanding of those substrates as they come with their own borrowed troubles each different from the other. This analysis here is not to fan support to the prevalent creationist young earth model. The Quran does not support a young ten or a hundred thousand (year) old earth concept; period. Even Noah's ark per the Quran is still antecedent to stone age dates of all human civilizations.[110] As this small band of people (believers) in the ark with Noah happen to father all nations of the world: *Quran 36:41 "And a sign for them is that We carried their forefathers in a laden ship."* The analyses here is to caution that we have got our fundamentals wrong in a way that we accept anything coming from mainstream science circles as true and continue to remain less privy to the truth regarding the Universe.

---

[110] Stone age: Dating this period is again a matter of debate. The Aad of Palmyra (Syria)—popularly known as the bride of the desert, make a spectacular stone age grandeur with -stone tools- discovered dating to around 7500 BC (Wikipedia). Obviously, the bride of the desert is a spectacle produce by a series of many generational feats. Considering this to be the oldest major human civilization on earth to have left wonderous signatures of their magnificence still thousands of years previous to them as these nations first sprung antecedent to this was Noah's appointment with God.

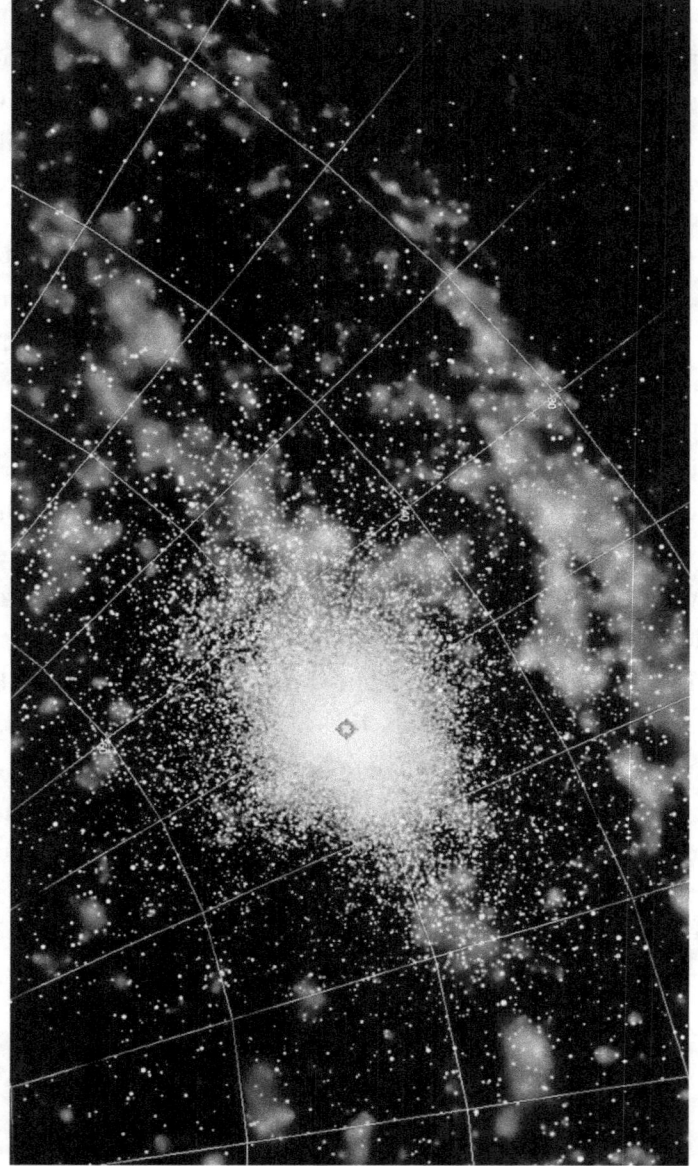

Figure 9 *Sector of Milky Way- Orion Arm.*

Orion spur- Sun is at the crosshair surrounded with high star density; It is evident that solar neighbourhood is packed with very rich and high-density stars. ***Figure 10*** *next, shows a zoomed view of this slide and see* ***Figure 12*** *for many different arms in Milky Way- Submitted via GAIA Sky software, version 1.0.4, 2016. Credits: GAIA Sky software.*

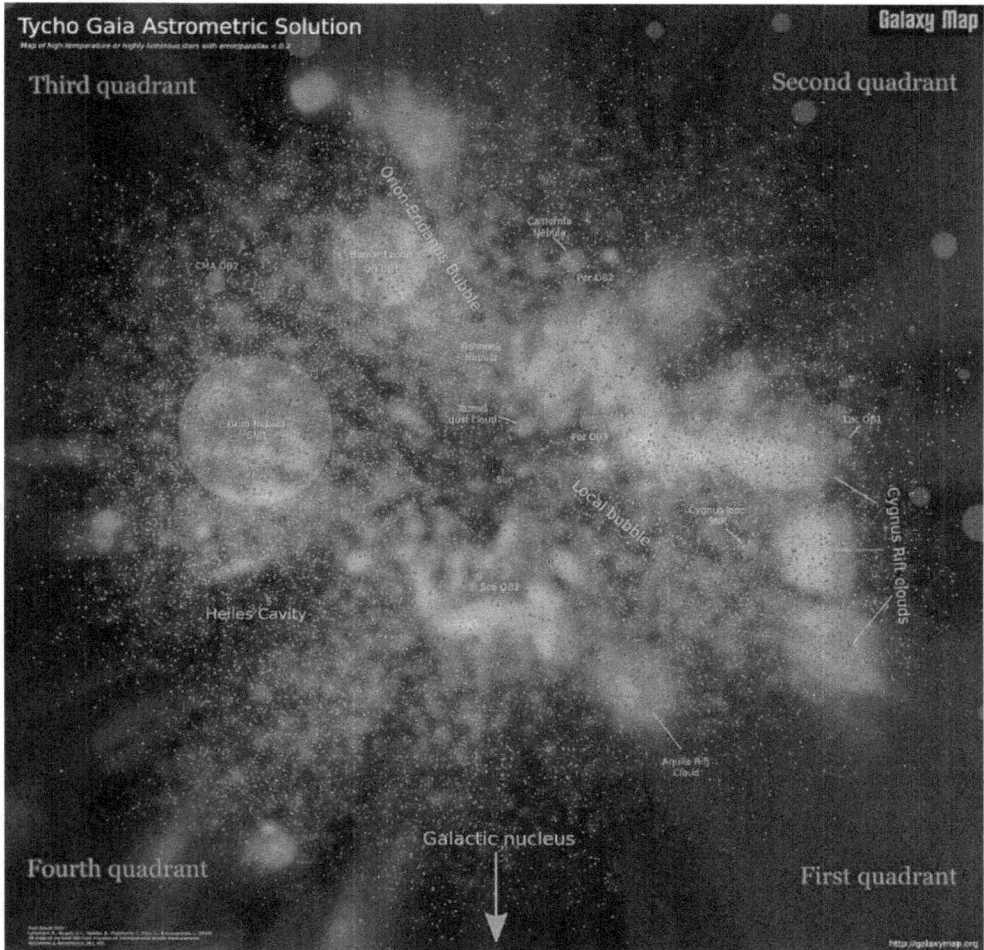

Figure 10 *First Cosmic Sky at a volume of 2600 ly.*

Tycho Gaia Astrometric Solutions (TGAS) high temperature/luminosity map- Sun is at the centre of frame- Submitted by Kevin Jardine on 16 September 2016. The data with error/parallax ratio < 0.2 goes out to about 700 parsecs or so. The map has a radius of 800 parsecs (2600 ly), which includes well all the Gould Belt. So, this map really shows all of what most astronomers would consider the solar neighbourhood. Image source: galaxymap.org/blog. Credits: Kevin Jardine.

Figure 11 *Sirius binary star system.*

Sirius A; White dwarf Sirius B seen towards left bottom corner; Realtime capture- Hubble Telescope; Image sourced: Wikimedia commons; Credits: NASA, ESA, H. Bond (STScI), and M. Barstow (University of Leicester).

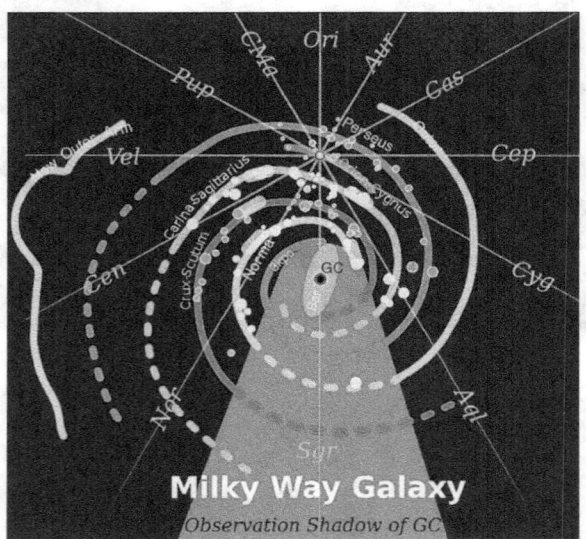

Figure 12 *Milky Way/Angular constellation vectors.*

A "God's view" map of Milky Way as seen from far Galactic North (in Coma Berenices). The star-like lines centre in a dot representing the position of Sun. The spokes of that "star" are marked with constellation abbreviations, "Cas" for "Cassiopeia", etc. The spiral arms are coloured differently to highlight what structure belongs to which arm. Image sourced: Wikimedia commons; Credits: Rursus.

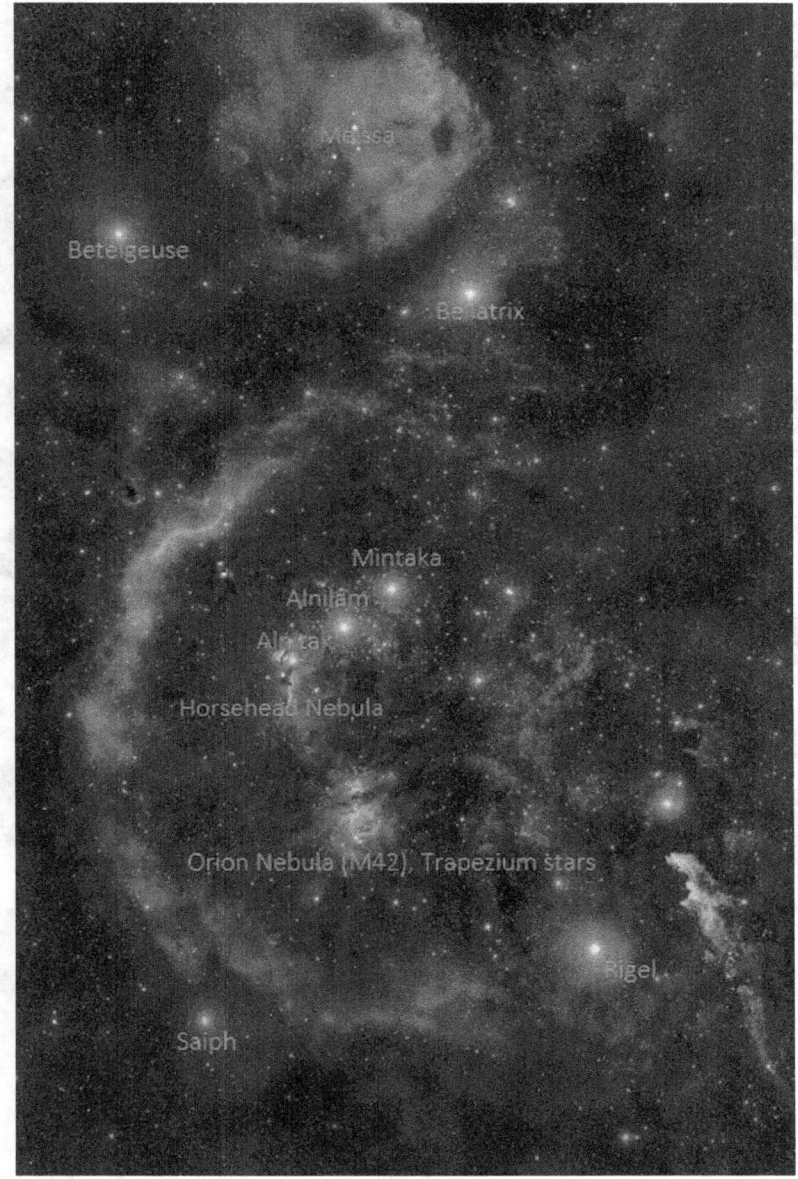

Figure 13 *Orion Molecular Cloud Complex.*
Betelgeuse and the dense nebulae of the Orion Molecular Cloud Complex; Galaxy: Milky Way;
Image sourced: Wikimedia commons; Credits: Rogelio Bernal Andreo.

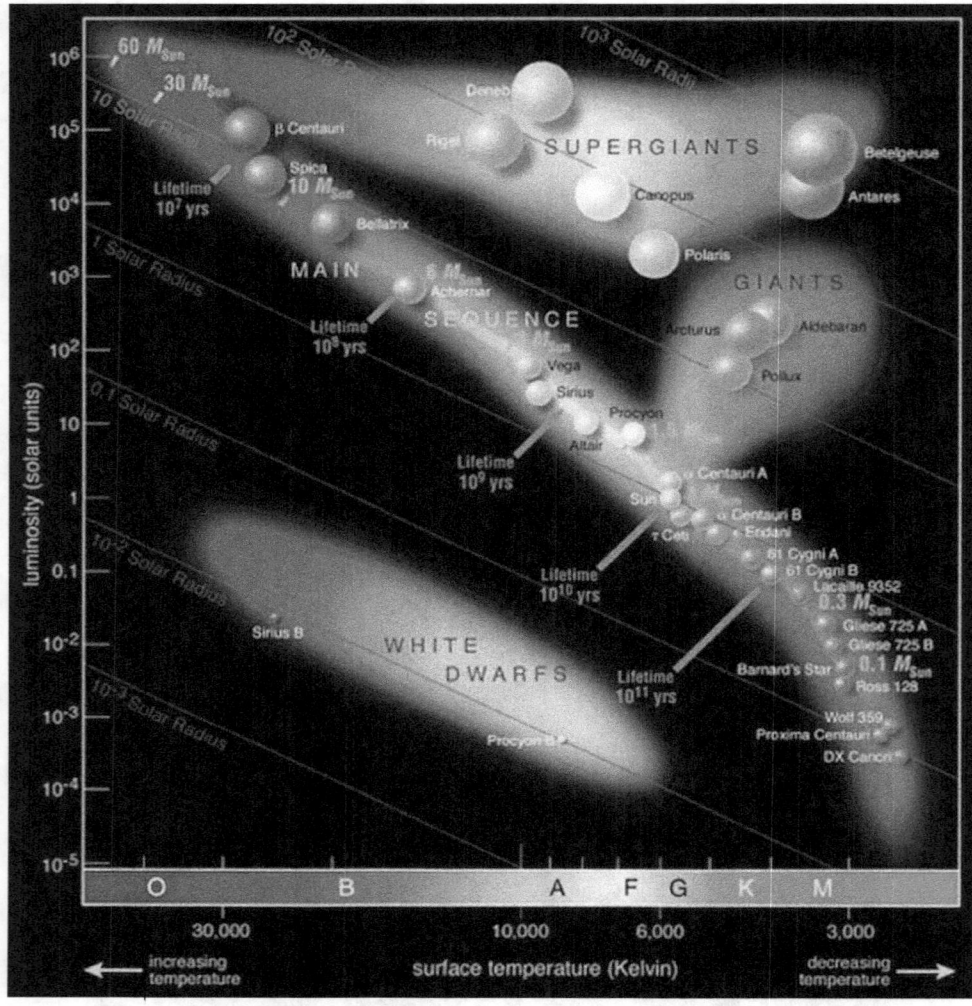

Figure 14 *Hertzsprung-Russel diagram.*

Hertzsprung-Russel diagram identifying Betelgeuse and well-known stars; Image sourced: Wikimedia commons; Credits: European Southern Observatory (ESO).

*Affluent Ramesses II was far from seeing assemblages of stars—
despite Egyptian heights, his bricks of mud crumbled felling him
from the sight of his Lord; Moses neither in the wilderness of Sinai
was granted vision but was turned down telling for the blind of an
eye yet have faith.*

Holy Quran is a revelation from the most merciful Lord who created planet earth and its skies in meticulous detail and has since established Himself upon His throne. Prior to this there was another act of Creation when Allah had created the major Cosmic skies which are beyond the seven earthly skies. It is amidst one of these that he placed the earthly skies and the solar system, a conclusion drawn from the verse 25:61 "...*(Allah) has placed in one such starry assemblage the sun and a lightsome moon.*" For ease of understanding and uniforming terminology the seven skies of earth are always referred herein as earthly skies (or minor skies) while the skies which are beyond will be referred as Cosmic skies (or major skies). Within the first of such Cosmic skies, in a grid, Allah placed our star and in diligence fashioned the planet earth to its finer details and made the moon and has since intricately imbedded mechanisms through it to keep His awareness of us and of everything surrounding us to the very details of our unmindful thoughts.

Quran 20:4-7 A revelation from Him who created the earth and the **greater Cosmic** skies. The Merciful established upon His Throne. To Him belongs what is in all vaults and that is on the earth and what is between them and what is under the soil. And if you speak aloud- then indeed, He knows the secrets and what is (even) more hidden.

The Arabic phrase السماوات العلى (as-samawati al-oulaa) translated as **greater Cosmic** skies in 20:4 is His exclusive artistry. All existences belong to Him alone, the ultimate Owner and Sovereign from Whom is hidden nothing—the all aware, the one Who never sleeps.

> Quran 2:255 Allah- there is no deity except Him, the Ever-Living, the Sustainer of (all) existence. Neither drowsiness overtakes Him nor sleep. To Him belongs whatever is in the (Cosmic) skies and whatever is on the earth. Who is it that can intercede with Him except by His permission? He knows what is (presently) before them and what will be after them, and they encompass not a thing of His knowledge except for what He wills. His **Kursi** extends over the (earthly) skies and the earth, and their **preservation** tires Him not. And He is the Most- High, the Most-Great.

The Arabic word كرسي (kursi) meaning a chair in verse 2:255 is in context of the central pew of Allah—which is told to extend over the earthly firmaments. And Allah is not wearied of guarding His creations in their existences from His kursi. The Arabic word وسع (wasiā) translated as 'extends', it does not imply to any comparative sizes, but it alludes to the status of its magnitude. It implies that by its natural disposition, the kursi of Allah extends above encompassing His creation. This concept is further discussed in section 6.9, of how Allah makes proportional approach toward the earth bearing His throne. The Arabic كرسي (kursi) meaning chair is the exemplification of Allah's pew upon His positional residence (upon His throne) whose Arabic is العرش (Al-Arsh). The word كرسي (kursi) comes in the Quran twice of which one mention is of Prophet Solomon's chair where he would seat himself upon his throne in his kingdom. The syntax specific to Allah's chair is talked of being expansive extending greatly beyond skies of the earth up to an unspecified variable, voiding the implicit of limiting its size or definition.

The Arabic العرش (Al-Arsh) meaning the throne is the exemplification of Allah's positional residence. The word comes in the Quran 29 times of which 21 are syntax specific to Allah's Arsh, the residential throne. The Quranic implied meanings surrounding this concept describes it as an unconstrained construct of what it is made of. Meaning it is not fixed in one spatial ordinate. And it is not described in a static comparison with anything else in the Cosmos unlike Kursi (chair) for which a baseline extension is given from beyond the earth and its skies. Instead, Al-Arsh (the throne) is talked of expressly concerning its positional movements beyond and within the realms of the universe.

Circumambulation is the hallmark in the artistry in space. Planets and non-planetary objects traverse around stars; stars traverse in constellations around their galactic centres. A group of Galaxies known as local galactic groups traverse together about their centre of mass. A bunch of these local galactic groups form superclusters. A group of such superclusters traverse around their centre of mass known as local superclusters- All in the observable and beyond universe making circumambulation. For believers, an act of worship instituted is to circumambulate the symbols of Allah on earth too. This is also found imbedded in ancient religions like the religion of river Indus people i.e. modern Hindus. However, in Islam circumambulation is still pristine to symbolise dedication only to the house of Allah, Kaba in Makkah. The realm above the all observable and beyond is where the throne of Allah resides. All that is in the universe and firmaments below, in the earthly skies, and beneath the soil belongs to Him. There is none who owns a thing in the realm of the Cosmos but Allah, the Lord of the worlds.

In popular science, a new theory is said to predict Cosmic sky beholding whole of the universe is to our observed universe as is this observed universe spread 46 billion light years in every direction from earth is to an atom.[111] In seventh century prose it was narrated by Prophet Muhammad that the proportions of successive Cosmic firmaments are in the likeness of a ring lost in a vast desert from each celestial space.

---

[111] Cosmic Journeys- The Age of Hubble by Thomas Lucas Productions, Inc.; Category: Science & Tech Documentary Nature, published 16 April 2015 on YouTube; time: 18:27-19:00 minutes.

6.1 PRIMORDIAL CREATION

> Quran 35:1 All praise is (due) to Allah- Originator of the skies[112] and the earth. (who) Made the Angels messengers possessing wing (pairs), two or three or four. He increases in their creation as He wills. Indeed, Allah is over all things competent.

> Quran 65:12 It is Allah who has created seven (Cosmic) skies and from the Earth similar number. (His) command descends through between them so you may know that Allah is over all things competent and that Allah has encompassed all things in knowledge.

Given the question of which is more plausible the Big Bang or the Darwinian evolution, majority consensus in my opinion would drift away from the messy pilot of Natural selection to lean on choosing the former. However, to our surprise, new science revelations annoying as ever, critique the famously propounded theory of Big Bang! It is argued that Big Bang theory conflicts with truth by modern scientific findings. So what outcome is probable if not for this theory? A range of issues surround it, from the Singularity improbability, to its difficulty to explain Galaxy formation, to the inference that the Cosmic background radiation isn't a product of Big Bang and the horizon difficulty. Further, the Big Bang as not being conductive to General Relativity, then its violation of laws of thermodynamics and the observed evidences of continuous matter generation such as the earth's, is already laying huge shackles of Iron around it.[113] In fact, the stance that there wasn't a time when there was nothingness in the realm of Cosmos is clearly a statement that went forth from God. This is telling that in the realm of the past, infinity existences were evident such as of Allah's eternal existence! Thus, Allah's creation act of originating creation in the grids of the universe from nothingness is alluded to, where by 'nothing' it is meant as a persona as in a time when the persona of that part of the universe wasn't there followed by a time when it came into being. The overarching Cosmic frontier where Allah's throne resides upon which He remains

---

[112] Sky or Skies in quoting verses sometimes are braced () and pointed alluding to a sky-type. Otherwise Allah's narrative encompasses entire creation from our reference as the skies of the universe down to the earth.

[113] Title search: 'Criticism to Universal Big Bang' on omicsgroup.org; Journal of Astrophysics & Aerospace Technology.

firmly established is far away from any detection to us. A domain which He has its shackles buried deep into the universe and very much down to the earth. In fact, a type of reach that pervades everything seen and unseen. Even a thought conceived at the quantum level is instantly brought to His awareness and He becomes cognizant of it **by knowledge**.

> Quran 45:23 Have you seen he who has taken as his god his (own) desire, and Allah has sent him astray **upon knowledge**. And has set a seal upon his hearing and his heart and put over his vision a veil? So, who will guide him after Allah? Then will you not be reminded?

He is Allah the master knower of everything within our dimensions and that of beyond, of knowledge known or unknown, of matters revealed aloud or concealed within, of intentions subtle or obvious, of secrets divulged or retained, of thoughts occurring deep within the minds. Often, misguidedly humans construe Almighty Lord to possess similar limitations like ourselves. Almighty Allah is above such lowly attributions. He is the source of origins and is not the result of any prior creation. As it is obvious with our simplest, sincere and mindful of thoughts.

> Quran 57:3-4 He (Allah) is the First and the Lasting, the Evident and the Immanent: and He has full knowledge of all things. He it is Who created the skies and the earth in six-days, and then is firmly established on the Throne. He knows what enters the earth and what comes forth out of it, what comes down from sky and what mounts up to it. And He is with you wheresoever you may be. And Allah sees well all that you do.

If we apparently look at things surrounding our lives our immediate self directs us to its maker like may be Apple Inc for its design of stately techy things. A similar reflection over the nature of our own selves and of the creation dispersed on earth and the universe, we find that such a direction is equally effective in favour of the ultimate Creator and the Fashioner Lord – concluding God's existence. This is akin to certain realisation at the time of resurrection where there wouldn't be the slightest mystery left in the face of emergent reality; that our knowledge of God's eternal and self-subsisting self will become clear and definite. And we will come to know that He as consciously felt previously is the subtle, the uncreated master of the day of Judgement.

Quran 36:51-52 And trumpet be sounded; then from their burials to their Lord they will be driven. They will say, "O woe to us! Who has raised us up from our depositories?" (It will be said), "This is per the promise of Most Merciful and the Messengers were true."

Here on earth it is a test, set out for assessment of belief, predicated on these true feelings about our creator's existence. Thus, His attributes describing some of His dispositions are beyond our comprehension. Also, as requests to sight of Him are denied, hence forth practicality is retained for this test to unfold. Therefore, submission to Him as realized by our natural inclinations to surrender to His Majesty only bears the final impetus for a successful outcome.

Quran 85:1 By the sky natured to possess بروجا (**burujann**).

As highlighted earlier the five spectra of بروجا (burujann) translated as **starry constructed mansions** talked in the Quran are:

1. Stars of the السماء الدنيا (as-samaa ad-dunya) which is of the first Cosmic sky.
2. Stars tucked in mansions بروجا (burujann) of a greater sky. Such as mansions of star arrangements like a galaxy. Situated in one of this mansions is our sun and the moon.
3. Galaxy clusters or the Local Galactic Group continuing in circumambulation around gravitational centres.
4. Most likely a group of galaxy clusters forming the Super Clusters such as the Virgo Supercluster, making the largest known material structures in the universe in circumambulation about the Throne of Allah.
5. And, finally the sky filled with star sites of potential quasars, blazars or black holes, referred by the plural بمواقع النجوم (bi mawaqhiā an-nujum) meaning stars setting sites.

All these five categories can be taken as the plural بروجا (burujann). Therefore, burujann wholesomely connotes to the many different spectra of star clusters and galaxy clusters spread out in the arms of galaxies and in the filaments in the universe. Thus, the observable intergalactic hub made of luminous filaments of galaxies outward in all directions forms a vast webbed structure which is to us the visible space of the Cosmos.

6.2 SOLAR NEIGHBOURHOOD

As discussed in section 5.2 the first in the seven Cosmic skies is the interstellar solar neighbourhood loaded with ammunition. This perceivably goes beyond the 15-light years (ly) boundary drafted for stars surrounding our sun in science diagrams (**Figure 15**). This idea of small volume of solar neighbourhood depicted in some science diagrams does great injustice to the actual composition of solar neighbourhood talked in the Quran and observed in the celestial space defining its volume to at least 3500 ly from observations.

> Quran 37:6 Indeed, We (Allah) have adorned the **first Cosmic sky** with an adornment of bodies in whorls.

> Quran 67:5 And We (Allah) have certainly beautified the **first Cosmic sky** with glowing lamps (bright stars) and We have made them as deterrents for Satan-kind and have prepared for them the punishment of the Blaze.

These set of verses estate the solar neighbourhood with an adornment of bodies in whorls, surmising that our sun is surrounded by a rich and dense whorl of stars and non-starry objects for deterrence. The scriptural elucidation here is towards the densely packed region of stars in the Orion Arm of Milky way, spread across a breadth of 3500 ly in volume whose arm stretches over a length of 10,000 ly— a zone marking of a respective sky for us. Having us reckoned at the centre stage in the solar neighbourhood in the Orion spur here, we are very much gripped by Allah for His assessment of us. And we are told that a similar architectural arrangement is made for the Paradise planet determined for beginning afterlife. Understandably a dark sky without the adornment of stars to beautify is clearly unpleasant for earth or for paradise planet.

Further, jinns in Paradise planet won't be of the hating kind to us and therefore with no more missions to rebel, stars will then serve purposes of a magnificent gleam and for navigation in Paradise for various kinds of beasts perhaps even for us. The righteous among both men and jinn will reside there for aye, while the inconsiderate would be in a continental sized pit, paying for what they earned for their rebel explorations. Stars, in that Cosmic sky would have still served the same exact purposes of enriching minerals and materials which they have demonstrably served here on earth. As stars are factories for manufacturing heavy elements. Some materials originate only in massively sized stars. These massive giants are foundries to produce materials which do not get generated in stars the size of our sun, such as Iron. The Quran tells of this fact that Iron found on earth was **sent down** from our respective first Cosmic sky.

> Quran 57:25 We (Allah) have sent our messengers with Clear Signs and sent down with them the Book and the Balance (of Right and Wrong), that men may stand forth in justice; and **We sent down Iron**, in which is (material for) mighty war, as well as many benefits for mankind, that by which Allah may test who will help Him and His messengers stealthily: For Allah is Full of Strength, Exalted in Might (and able to enforce His Will).

It is not expressly told in the Quran of what else resides in the wider universe of other constellations within Milky Way or in other galaxies other than information regarding beings of special creation in Paradise planet and findings of similar flora and fauna as experienced here on earth to eat such as flesh of fowls, figs, olives, pomegranates, grapevines, stalks of dates, spathes of bananas, and for decor materials such as elegant cushions of silks, precious stones, glass crystals, Iron, brass & bronze, gold, and silver and minerals for wholesome enrichment: *Quran 47:15 "(Here is) a Parable of the Garden which the righteous are promised: in it are rivers of water incorruptible; rivers of milk of which the taste never changes; rivers of wine, a joy to those who drink; and rivers of honey pure and clear. In it there are for them all kinds of fruits; and Grace from their Lord. (Can those in such Bliss) be compared to such as shall dwell for ever in the Fire, and be given, to drink, boiling water, so that it cuts up their bowels?"* Such a detail anyway isn't a necessary info for us to have knowledge of, or for assenting to excellences in our respective missions here on earth to worship Allah.[114] However, as the past and future belongs to Allah, for He is the originator of creations as He pleases, He has informed us of what he deemed fit to inform concerning our future abode in another brilliant constellation, perhaps, maybe within the realms of Milky Way itself.

Where it comes to Astronomers' anciently drafted star classifications into constellations which are divided into angular sectors from the earth's observation

---

[114] Principle component in worship of Allah is to provide care and wellbeing for all to impress upon our Lord, this is deemed in the Quran as the required righteousness.

reference. The Quranic description of a constellation is inclusive of only those stars which are within the realms of solar neighbourhood. This is indicated in yet another place in *Quran 15:16-18 "And We have placed in the sky **starry constructed mansions** and have beautified it for the observers. And We have protected it from every devil expelled. Except one who steals a hearing and is pursued by a clear meteor (burning flame)."* The Arabic word بروجا (burujann) translated as starry constructed mansions in this verse and having it "beautified for the observes" in conjunction with another purpose of protection "from every devil expelled" connotes this boundary to be within solar neighbourhood. It is therefore apt to say that the set of verses here accord for their fitting in this definition of solar neighbourhood. The talked constellations here are well within the reaches of jinn forces to pry. Constellations of the far regions of solar neighbourhood and belonging to numerous starry clusters of the various arms of Milky Way and beyond belonging to separate galaxies are clearly out of reach for any jinn's effort to venture out see section 7.4 for details.

6.3 GALAXY ASSEMBLIES

Quran 25:61 Blessed! is He (Allah) who has placed in the sky بروجا (**burujann**) and placed therein an anchor (sun) and a lightsome moon.

Allah has placed in the sky numerous constellations and in one among them is located our sun and the moon. Just in the last section we read a different syntax for بروجا (burujann) where it was described as stars of the near world Cosmic sky i.e. solar neighbourhood or **first cosmic sky**. Verse 25:61 here presents a greater context stating that our solar neighbourhood is one such starry-constructed site of a larger sky. Numerous such starry-formations or numerous such first Cosmic skies like our solar neighbourhood makeup for the star concentrations in the arms of the Milky Way galaxy (**Figure 16**). It entails that the **second Cosmic sky** is the Milky Way and as there are multitudes of such galaxies, each of them constitute a respective second cosmic sky for their celestial estates. In one estimate the Milky Way is said to have a mass of 100 billion solar masses, so it is easiest said to have 100 billion stars accounting for the stars that would be bigger or smaller than our sun and averaging them out. Other mass estimates of Milky Way bring the number up to 400 billion stars, whichever is true and whatever the case we are seeing the astronomical structure of heavens being beautifully laid out before us from the Quran.

6.4 LOCAL GALACTIC GROUPS AND LOCAL SUPERCLUSTERS

The local group of galaxies see **Figure 17** is the galaxy group that includes the Milky Way. It comprises of fifty plus galaxies, most of which are dwarf galaxies. Milky Way and Andromeda make the prominent members in this group. It has a diameter of 10 Million ly (3.1 Million parsecs (pc)) and is spoken of as containing poor and irregular cluster of galaxies. Poor clusters it is said may contain only a few dozen galaxies, as compared to rich clusters which have hundreds or even thousands of galaxies. This local group is near the outskirts of a much bigger cluster (also known as the Virgo supercluster), which has more than a million galaxies in it and is more than a hundred million light years across.

As mentioned earlier, Prophet Muhammad's analogy of the Cosmic skies was like a tiny ring lost in a vast desert where if the ring is taken as the sky then the desert would be its next encompassing sky. From modern astronomy corroborated data, we can therefore construct the frontiers of these skies by extrapolating observational evidences to distinguish them. By the Astronomers' draft, Milky Way is classified as part of Sagittarius constellation and the prominent Andromeda in the Local Galactic Group is consolidated as part of the Andromeda constellation. But, by the Islamic narrative a more accurate method of cluster-based collection of galaxies dictated by the golden rule of circumambulation is drawn. The lack of insights into mapping circumnutating of galaxy groups about the rotational axes in the next celestial connect of the superclusters and within the disparate local units, blocks the needed information for classifying them uniquely. In fact, correct knowledge of بروجا (burujann) starry constructions in clusters which undertake full circumambulation with their next celestial connect or gravity centres graduates to them being a constituent for any unique Cosmic sky.

Going by this view all galaxies in the Local Galactic Group consisting of the two big galaxies i.e. Milky Way and Andromeda and additionally fifty plus galaxies with the gravitational centre located somewhere between the two big galaxies may be considered as a unique Cosmic sky if full circumambulation is thought to occur. Also considering that our local group moves in tandem with other local galactic groups in clusters around the next celestial rotational axis in another level of circumambulation then such a cluster uniquely classifies as the next Cosmic sky in celestial space. Therefore, the Local Galactic Groups makeup the

third Cosmic skies and the Local Superclusters makeup the respective **fourth Cosmic skies.**

6.5 SUPERCLUSTERS

A supercluster is a large group of smaller galaxy clusters or a gathering of many Local Galactic Groups (see **Figure 18**). It is among the largest-known material structures in the universe. Tracing Milky Way, it is found to be part of the Local Galactic Group which in turn is part of the Virgo cluster which is a part of an even more staggeringly huge structure called Virgo Supercluster, spanning over a hundred million ly. Virgo supercluster is discovered to be part of Laniakea (Hawaiian for "immeasurable heaven") trailing like an appendage to it. Laniakea spans over 500 million ly (see

Figure 19). The most massive superclusters of Laniakea are Virgo, Hydra, Centaurus, Abell groups, Fornax, Eridanus and Norma. Each of these superclusters are potentially the previously discussed fourth Cosmic skies attracted about the central gravitation near Norma called the great attractor. Thus, Laniakea emerges as the **fifth Cosmic sky** if it can be said accurately of our present understanding.

Therefore, the first, second, third, fourth and fifth Cosmic skies are respective skies in circumambulation about their next gravitational centre. The next celestial connect of Laniakea is not known in our current scientific endeavours. As the component clusters of Laniakea are noted for Hubble flow, for fulfilment, I would assume it to be in obedience to Allah's making of the expanding sky discussed in section 6.7. This kind of subset classification of galaxy clusters also clarifies the definition for local directing of dooms for probable worlds resident in them or perhaps which may become plausible in the future. In our case therefore it derives that doom for us will be effective within our respective realms of our celestial space.

Excerpts from Wikipedia: The biggest cluster in the observable universe is called the Great Attractor. Its gravity is so strong that the Local Supercluster, including the Milky Way, is moving in a direction towards it at a rate of several hundred km per second. The biggest supercluster outside of the local universe is the Perseus–Pegasus Filament. It contains the Perseus supercluster and it spans about a billion light years, making it one of the largest known structures in the universe. Research has tried to understand the way superclusters are arranged in space. Maps are used to display the positions of 1.6 million galaxies. Three-dimensional maps are used to further understand the positions of these superclusters. To map them three-dimensionally, the position of the galaxy in the sky as well as the galaxy's redshift are used for calculation. The galaxy's redshift is used with Hubble's law to determine its position in three-dimensional space. The existence of superclusters indicates that the galaxies in the Universe are not uniformly distributed; most of them are drawn together in groups and clusters, with groups containing up to some dozens of galaxies and clusters up to several thousand galaxies. Those groups and clusters and additional isolated galaxies in turn form even larger structures called superclusters. Superclusters form massive structures of galaxies, called "filaments", "supercluster complexes", "walls" or "sheets", that may span between several hundred million light years to 10 billion light years, covering more than 5% of the observable universe. These are the largest known structures to date. Observations of superclusters can give information about the initial condition of the universe, when these superclusters were created. The directions of the rotational axes of galaxies within superclusters may also give insight and information into the early formation process of galaxies in the history of the Universe.

6.6 OBSERVABLE UNIVERSE

As unfathomable as it may sound, when we imagine the Universe as one single entity with its collection of the most massive superclusters arranged such as Hercules-Corona Borealis Great Wall[115]—the first structure to exceed the size of 10 billion light years, Shapely supercluster among the moderate of sizes to which the Great Attractor with its components of Laniakea is counted drifting to and an additional list of at least forty on Wikipedia such as Lukash LQG 10 structure among the smallest of sizes, with large spaces of voids between them, arranging into greater formations called filaments, walls or sheets, we come across definitions which are so massive that the literature falls short to explain. We do not have the words to address these intricate findings. These are none other than collections of the multitudes of mighty **fifth cosmic skies** in respective senses. Within these are sites heavily needle-worked to annihilate baryonic starry-materials such as black holes advancing into quasars and blazars which are deductive of the below verses:

Quran 85:1 By the **sky** natured to possess بروجا (**burujann**).

The Arabic word برج (burj) meaning a tower is a singular form of بروجا (burujann) used here, to describe mansions of stars as an attribute of this sky. برج (burj) literally means a building or a tower with several units in them such as the Orion constellation, Canis Majoris, Ursa Major etc. each containing multitudes of stars. Several of these برج (burj) make up بروجا (burujann) and in the greater celestial connect they then graduate to become a singular assembly once again defined as برج (burj). This is because the word السماء (as-samaa) meaning the sky encloses several of these constellations as a sole firmament. In the realm of major Cosmic skies, the Arabic phrase السماء الدنيا (as-samaa ad-dunya) meaning 'sky of this world' defines the first Cosmic sky as the celestial space of solar neighbourhood to be a unique and a respective sky. Therefore, all unique star assemblages akin to solar neighbourhood are then referred to as a plural بروجا (burujann) making the celestial space of Milky Way. Milky Way again in the next celestial arena is a برج (burj) i.e. a singular firmament, then combining with several similar firmaments to form بروجا (burujann) called Local Galactic Group. Which again successively patterns up, principled by the characteristic potential of full-

---

[115] Discovered via gamma-ray burst mapping, Wikipedia.

circumambulation into greater firmaments of successive skies known as السماوات
العلى (as-samawati al-oulaa) translated as **greater Cosmic skies** as cited in the verse:
*Quran 20:4-5 "A revelation from Him who created the earth and the **greater Cosmic**
skies. The Merciful established upon His Throne."*

> Quran 56:75-76 Furthermore I call to witness the **setting**
> **locations of the Stars**. And that is indeed a mighty adjuration if
> you but knew.

The **sky** mentioned in 85:1 is connoted to mean the observable and beyond
observable universe. The overarching Universe (see **Figure 20**) with its capacity to
possess hosts of star setting sites arranged in tendrils of the mighty architecture
of the collection of fifth Cosmic skies is the common expanding universe. In verses
56:75-76 referred herein speaking of 'setting locations of stars', each galactic
centre or a more appropriate idea is that of a devouring Galactic centre called as
quasar serves for dooming of star clusters. They are also called as blackholes while
they still behave as infants, hardly having begun feeding on its starry materials.
The Arabic phrase بمواقع النجوم (bi mawaqhiā an-nujum) meaning 'stars setting
sites' in verse 56:75 is invoked by Allah as the mighty adjuration, for the aura of
the Milky Way galactic centre which perhaps has not yet begun to feed on its
starry material (see **Figure 21**). When Allah magnifies this oath as the mighty
adjuration, it is not akin to exploding of a star in its positional locus like when a
main sequence turns into a dwarf or exploding of it into a supernova. But it is the
annihilation of clusters of stars or mansions of starry constellations which is
indicated in the verse using a plural form in the syntax بمواقع النجوم (bi mawaqhiā
an-nujum) translated as sites in the wider expanse of the universe for clusters of
stars annihilation. Such sites other than known as black holes are popularly called
quasars and blazars especially in their voracious feeding sequences, emitting huge
X-ray jets extending at least a Million ly from the quasar. Eventually post our
judgement and the celestial transfer solar neighbourhood will be led to such an
end in the galactic centre.

> Quran 56:77-79 That this is indeed the Quran Most
> Honourable. In a Book, well-guarded (in the firmament with
> Angels). Which none shall touch but those who are made for it
> (Angels).

The well-guarded book or entity referred in verse 56:78 refers to the symbolic law book of Allah containing knowledge of natural dispositions of all creations ever created or imminently forthcoming and then excerpts from this are sent to respective celestial spaces. This is the fundamental source of inscription and guidelines given to Angels to perform, which documents the purposes in God's creations and the ends of His objectives. Parts from it were time and again channelled to generations of peoples in understandable syllables of vernacular languages from the distant past of human beginnings. The Quran making the final deliverances to Prophet Muhammad, is one among an extract from it, thus Prophet Muhammad delivered it to mankind and took a popular public testimony for it in his most remembered and commemorated farewell speech.

> Quran 56:80-82 A Revelation from the Lord of the Worlds. Is it such a Message that you would hold in light esteem? And have you made it your livelihood that you should declare it false?

Death is therefore the determinant to demonstrate Allah's majestic reach over His creation. And the total annihilation of the obsolete is done via transporting it to the quasars. An array of such sites is arranged in filaments with enough potential to gravitate dooms in respective grids of the galaxy clusters. As is observable in nature even with vegetation, it is this attribute of Allah that He brings things into creation then fulfilling their purposes leads them to their demise and again regenerates from it entirely new creations. Remarking by several such similitudes He proves to mankind that their resurrection after death is a binding promise upon Him.

Since the 1960s, astronomers have been studying bright beacons of light that shine from the centres of distant galaxies. A quasar's power source they found, is a black hole that has grown to millions, even billions of times the mass of our own Sun. The thinking is, that magnetic fields beam off a disc of matter, then spiralling into the black hole, drawn by its extreme gravity. These fields channel a portion of the inflowing matter out into powerful particle beams. The jet is part of a larger rush of matter away from the black hole. You can see it in a large spiral galaxy called NGC 3783, 30 million light years from Earth. Astronomers use the Very Large Telescope array in Chile's Atacama Desert to peer into the core of this galaxy to study the environment of a super massive black hole. From a disc of matter flowing into the black hole, intense radiation had created a dusty wind that is moving up and away from the black hole. The source of the dust is likely generations of giant stars that lived and died in the galaxy's central region. Black hole winds are now thought to have had a major impact on the universe at large. You can see it in a simulation of early cosmic evolution. From starting point 12 million years after time zero, this computer simulation shows the evolution of a cosmic patch, some 350 million light years across. Not everything is known about what happened including the nature of a dominant substance known as dark matter. Within the volume of this simulation, tens of thousands of galaxies take shape all along the strands of the cosmic web. Where the filaments intersect, something happens that will affect the character of every galaxy, star and planet. Gigantic bubbles of hot gas begin to expand outward. They form when matter flows into large central galaxies and gets blasted out by super massive black holes, lurking in their cores. In time, these hot bubbles push out well beyond the central galaxies. This has the effect of limiting the amount of gas that can fall into central regions, allowing smaller galaxies like ours to form on the periphery. The bubbles spread huge volumes of gas and dust created in earlier generations of stars. Galaxies all around the universe bear witness to this dusty legacy. ... Here, above the plane of the galaxy, just 400 light years from Earth is a dense irregular cloud of dust called Lupus 4. Our Sun, our solar system likely formed in a cloud like this.[116]

---

[116] Cosmic Journeys- The Age of Hubble by Thomas Lucas Productions, Inc.; published 16 April 2015 on YouTube, Time:22:48 onwards.

6.7 EXPANDING SKY AND THE INVISIBLE MASS

The number of superclusters in the observable universe is roughly estimated to be 10 million. Galaxies are grouped into clusters instead of being dispersed randomly. Clusters of galaxies in turn, are grouped together to form superclusters. Unlike clusters, most superclusters are not bound together by gravity. The component clusters and superclusters are thought to generally shift away from each other due to the Hubble flow. Hubble's law is considered the first observational basis for the expansion of the universe and today serves as one of the pieces of evidence most often cited in support of the Big Bang model. The motion of astronomical objects due solely to this expansion is known as the Hubble flow. Objects observed in deep space (extragalactic space, 10 megaparsecs (Mpc) or more) are found to have a Doppler shift interpretable as relative velocity away from Earth. This "Doppler shift measured velocity" of various galaxies receding from the Earth, is approximately proportional to their distance from the Earth for galaxies up to a few hundred megaparsecs away.[117]

> Quran 51:47 And the sky We (Allah) constructed with power, and indeed, We are (its) **expander**.

Expanding universe is plainly communicated. I have therefore considered this verse to list the **sixth Cosmic sky**—the sky of an expanding observable and beyond universe. A sky that is invoked for expansion causing the redshift in the Electromagnetic (EM) spectrum captured of the light coming from distant stars and galaxies from the fringe units of Superclusters. As astronomers began looking at exploding supernovae that trickled into our telescopes these cosmic beacons were measured for their brightness with another measure; how far their light had shifted to the red. The larger the shift, the more the universe had expanded. As highlighted in section 6.5 the component clusters forming Superclusters are generally shifting away from each other due to the Hubble flow. At this expanse of the universe, the red shift in the light spectrum of those fast displacing stars is noted. Luminous filaments of Local Galactic Groups drift away indicating an

---

[117] Internet search result: The Doppler effect (or the Doppler shift) is the change in frequency or wavelength of a wave for an observer who is moving relative to the wave source. If you are approaching a noise it gets louder as you close in and if you turn away it subsides, which can be understood as small wavelengths with high frequency and long wavelengths with low frequency.

expanding universe. An observation far removed from concepts of premodern astronomy let alone middle ages!

The Arabic phrase والسماء بنيناها بأيد وإنا لموسعون (wa as-samai banainaha bi-ayidinn wa innaa la-musiāuoon) in literal sense translates as follows: And the sky We (Allah) constructed with our hands and indeed, We (Allah) are (its) expander. The active participle لموسعون (la-musiāuoon) meaning I am the expander comes from the same word وسع (wasiā) translated as 'extends' that is discussed earlier in section 6. In the surah named Adh-Dhariyaat (chapter 51: The Winnowing) the Quran answers enquires of Makkan disbelievers concerning earthly punishments and doomsday. The judgement scene assumes from verse 51:5 very early on in this context, as it talks about the Cosmic pathways *Quran 51:7 "By the sky containing (crocheted) pathways"*, indicating space-time for the eventual doomsday event to unfold. This contextual verse is developed recounting the earthly fates of previous peoples as highlighted ahead in section 7.5. Then this narrative concludes context of final impending judgement in a verse elucidating expansion of the great sky—the universe, *Quran 51:47 "And the sky We (Allah) constructed with power, and indeed, We are (its) expander."* In verses after this verse a new theme is brought forward to recount the favours of Allah to mankind in making our earth soft-spread for us to inherit which contains all sorts of living species in pairs such that we may know that Allah is Kind. And He only sends Messengers to warn mankind of His impending acknowledgement. This Surah mentions two unique Cosmic skies- the Expanding sky and the Cosmic sky full of pathways, which are counted in the list as **sixth & seventh major skies**, a pair of miraculous mentions indeed!

The force causing space to speed up is still unknown. Scientists refer to it as dark energy and it may be all around us. Far from being empty, the vacuum of space is filled with energy, that constantly wells up in the form of particles of opposite charge, matter and antimatter.

> Quran 7:185-188 Don't they reflect in the government of the (Cosmic) skies and the earth and in all that Allah has created? That it may well be that their term is nigh, drawing to an end? In what fancies after this (Quran) will they then believe?

186. To, such as Allah rejects from His guidance, there can be no guide: He will leave them in their trespasses, wandering in distraction.

187. They ask you about the (final) Hour- when will be its launching? Say (O' Muhammad), "The knowledge thereof is with my Lord (alone): None but He can reveal as to when it will occur. **It is burdened heavily in the heavens and the earth**. It won't be, but of a sudden released upon you." They ask you as if you are familiar with it. Say: "The knowledge thereof is with Allah (alone), but most men know not."

188. Say: "I have no power over any good or harm to myself except as Allah wills. If I had knowledge of the unseen, I should have multiplied all good, and no evil should have touched me. I am not except a warner and a bringer of good tidings to a people who believe."

The set of verses quoted here is in elucidation of verse 51:47 talking about the expanding sky. These verses 7:185-188 from surah Al-Araaf (The Heights) are exquisite in telling that the release of doomsday is tantamount to a surprise and clearly a sudden resulting phenomenon without the trace of a foreseeable clue, coming from an unseen dimension. Once it commences there is no averting of it. This suddenness is attributed to an impending displacement that will be effective in the sky loaded with heavy-mass burdening. The Arabic word ثقلت (thaqhulat) meaning heaviness and mass is a scalar quantity, qualifying the measure of weight. Hence, the translation of **heavy burdens loaded in the universe** is connoted. In present Astronomical understanding the description of dark-matter and dark-energy is an analogous concept. As and when it will be pocketed out from a region, astronomical objects withheld within its sphere lose its stability and tellingly fall off toward the shifted centre of gravity into quasars akin to snapping of a system resulting from broken balance.

When it comes to dark energy more is unknown than what is known. The dark energy and matter descriptive of Quran 7:187 could be considered an appendage in elucidating the expanding universe. The dark energy and dark matter are presently thought of to make up 95% of volume in the observable universe. Astronomers know that this elusive matter is distinct from baryonic matter (ordinary matter such as protons and neutrons) which otherwise explains many puzzling Astronomical observations. It is indeed astonishing that modern astronomy conforms with the Quran for its informing us about this invisible heavy burden permeating through celestial space and loading the universe and the earth with it. But again, it is also not astonishing, for it is the book from its creator.

Astronomers say: "We know how much dark energy there is because we know how it affects the universe's expansion. Other than that, it is a complete mystery. But it is an important mystery. It turns out that roughly 68% of the universe is dark energy. Dark matter makes up about 27%. The rest- everything on earth, everything ever observed with all our instruments, all normal matter- adds up to less than 5% of the universe. Come to think of it, maybe it shouldn't be called "normal" matter at all, since it is such a small fraction of the universe."

6.8   COSMIC HIGHWAYS

> Quran 51:5-7 Indeed, what you are promised is true. And
> indeed, the Judgement is to occur. By **the sky containing
> (crocheted) pathways**.

It may not be obvious yet, as nothing about pathways is known to exist
but only speculated from studying the many extensive researches launched into
celestial space. The theory of wormholes where space boundaries are warped into
loops to travel across space-time is a very popular notion in astronomy.
Wormholes are predicted by the theory of general relativity, but none have been
discovered till date.

The Arabic phrase والسماء ذات الحبك (wa as-samai zatil hubuk) mentioned in the
verse 51:7 translates as -and the sky containing (crocheted) pathways-. As seen
in this, the Quran emphatically reveals a characteristic nature of the sky to
contain pathways. The word حبك (hubuk) meaning crochet, knit or weave is an
elegant description of how jampacked the sky is filled with these highways. The
sky of pathways extends from the throne of Allah and pervades down to earth
and to every corner site of the Universe. Angels cruise at targets as and when
required using these pathways. It is told that not just Angels but the words of
supplications uttered by believers and the deeds of their actions performed to
impress Allah also travel to Him through these crocheted tunnels: *Quran 35:10
"Whoever desires honour (they may realise)- to Allah belongs all honour. To Him
ascends good speech, and righteous actions rise unto Him..."* These ascended deeds
are recorded with Allah despite His prior knowledge of it only to be presented as
evidence for honouring of believers on judgement day.

> Quran 70:4 The Angels and the Spirit (Arch Angel Gabriel)
> ascend to Him (Allah's throne) in a Day which has a measure
> equal to fifty thousand years.

It is alluded that Angels staffed in first Cosmic sky in the Orion house consume
travel-time to Allah in a time frame which is in our reckoning a measure of fifty
thousand-years. In another verse the Quran says:

> Quran 32:5 He arranges matters from the (earthly) sky to the
> earth; then its (reporting) will ascend to Him in a Day, the
> extent of which is a thousand years of those which you count.

Since the creation of planet earth and its system of skies, Allah assumed to His throne from where He directs matters concerning this world. In the context of a preceding verse *Quran 32:4 "It is Allah who created the skies and the earth and whatever is between them in six days; then He established Himself above the Throne. You have not besides Him any protector or any intercessor; so, will you not be reminded?"* the interpretation of earth and its skies is brought to stage telling of recording its various states for responses to realising orders of Allah taking place. All Matters from earthly skies down to earth are regulated and then its reporting reaches its place of inscriptions with Allah via the Cosmic pathways (channels) in a time span of one thousand-years of our reckoning. Thus, the reports are posted to scribes with Allah by Angel- scribes from the earth, who transmit recordings faster than speeds they are themselves given to cruise. This delivery reaches its destination in a day reckoned in our time-frame of one thousand years' time, which is fifty times faster than the speed of Angels themselves travelling to reach God Almighty as revealed in the verse 70:4 cited before.

> Quran 22:47 And they (disbelievers) urge you to hasten the punishment. But Allah will never fail in His promise. And indeed, a day with your Lord is like a thousand years of those which you count.

Further, it is indicated that the gravity at the throne of Allah shrinks time for a day with a thousand years of magnitude in our reckoning. Which means a thousand years of human activity on planet earth corresponds to a unit measure in Allah's assembly. A day with God should not be misunderstood for days, nights of sun's splendour or moon's lightsomeness and years like here on planet earth but it is a measure explaining time itself with respect to our times on earth. This lets us know that plans of God are timed for the futuristic thousand years of our reckoning. In which He deems whatever fit for us for the next thousand years to settle our performance as civilizations and tribes and then He brings about His arguments for concluding determinisms [118] on nations. Therefore, Allah ever admonishes believers to remain patient and persevering regardless of enticements from hypocrites and oppressors determined to showcase mischief.

---

[118] Determinism: Also a decree of God to interfere into human affairs at the level of wider population. It mostly concludes in death & destruction and can also result in suffering and losses.

Quran 22:40 ...and if Allah did not check the nations, some by means of others, there would have been plundering of monasteries, churches, synagogues, and mosques in which the name of Allah is much mentioned. And '*Allah will surely support those who support Him*'. Indeed, Allah is Powerful and Exalted in Might.

The indiscriminate Almighty shines again as ever with aid announced to those among faithfuls who advance the cause of God, above all, regardless of any subscribed religion wherein people make a monotonic claim upon God. If clear faith in Almighty God alone is kept following it up with righteous conduct, then God will certainly aid them above others for He is more inclined to see fairness and justice from a people more than their verbal deliverances. As seen the recipients of His books in the past have in time been uprooted as they inclined to fellowships of desires therefore God's aid is indiscriminate as told above.

6.9 ALLAH'S ABODE

> Quran 67:16-17 Do you feel secure that He who is in the sky would not cause the earth to swallow you and suddenly it would sway? Or do you feel secure that He who is in the sky would not send against you a violent tornado? Then you would know how (severe) was My warning.

It is told that Allah's reach is ever near despite His Al-Arsh (throne) being situated across celestial firmaments. This throne of Allah is unique. It is the vault from where He rules encompassing all other celestial firmaments of the far reaching and beyond observable universe. Allah's reachability and might pervades everything in the fabric of the Cosmos.[119] Therefore, His orders are swift brooking no delay. All records are then logged into the books of evidences to which are appointed scribes who do not but remit every little thing to be produced as evidence before the deniers on Judgement day, a day when all histories of deeds will be laid open manifested before man. Man, then will be left dumbfounded to acknowledge his fate far from arguing any further with the Lord of the worlds.

> Quran 11:7 And it is He (Allah) who created the skies and the earth in six days. And His Throne had been upon water- that He might test you as to which of you is best in conduct. But if you (O Muhammad) say, "Indeed, you are resurrected after death," those who disbelieve will surely say, "This is nothing but drivels."

A startling detail is communicated in this verse here. Until recently scientists had asserted non-availability of water inferring its findings would be signatures of life wherever it is found extra-terrestrially. So, space missions were on principal to finding water outside planet earth's domain. But to their surprise water is found readily within our solar system in some planets, in some of their moons, in large number of space objects such as asteroids and in the expansive regions of the Cosmos. The above verse telling of Allah's Al-Arsh (the throne) had been upon water at the creation site is simply unthinkable, a detail explicitly communicated 14 centuries ago. Its significance can only be known and praised in the highly

---

[119] Quran 56:83-87 Then why do you not (intervene) when (the life of dying) reaches the throat. And you the while (sit) looking on. But We are **nearer to him** than you, but you see not (angels of God causing death).

knowledgeable circles. In the creation of planets water plays a crucial role, without water silicates would not gel as needed to form masses between colliding particles. Allah's throne is greatly sized, its stationing upon water would mean a very wide-field of water, not Amazon, not even the width of the Pacific Ocean but perhaps much, much, larger than it. In celestial space exploding stars are cited for formation of huge volumes of water as wide as their pattern of explosion itself. Even a protostar (new born star) is able to generate water as great as 100 million times the rate of Amazon flowing.

*Dying stars are not the only source of water. 750 light years from Earth, a protostar called L1448MM is shooting matter out in high speed jets containing oxygen and hydrogen. When these molecules reach a cooler environment, they form water, at the rate of **100 million times the flow of the Amazon River.** Water then becomes part of the dynamics of a **solar system's birth** ...*[120]

Indeed! This Quran could never be the work of a human: *Quran 17:88 "Say (O' Muhammad), 'Even if mankind and the jinn gathered in order to produce the like of this Quran, they could not produce the like of it, despite backing up each other in support'."*

Besides the problems in understanding of how gravity hems the matter to swirl together initiating the process of creation, which Allah simply states in His words: *Quran 35:1 "All praise is (due) to Allah- Originator of the skies and the earth...."* The precursors to this event are not yet clear. Such as the first realisation of hydrogen in celestial regions from thin space or the forming of filaments of the visible universe which is also understood as the formation of first baryonic matter. Our story begins from the gas and dust clouds that are created and recreated from galactic centres and exploding stars. The winds of the galactic centres shape these gas and dusts clouds in their peripheries of influence. The gas cloud collapsing in gravity gives birth to a star, which attracts the matter surrounding the protostar to encounter more gravity contrails. As the matter flows into it to settle, a series of grooves gorge in giving rise to new gravity centres where the planetary matter begins accretion to form many protoplanetary discs wherein planets form. Along the way, water acts like a collating agent, gluing

---

[120] Cosmic Journeys- The Age of Hubble by Thomas Lucas Productions, Inc.; published 16 April 2015 on YouTube, Time:36:20 onwards.

dust grains to form larger bodies, from rocks to comets, asteroids and fiery planets and nonetheless blue life-giving jewels not any other than earth yet known to science. These are some of our simplistic astronomical observations. Does this humble us or provide us with sanctions to disregard God?

Our new understanding of phenomenon unfolding in celestial space has given us the wisdom to now comprehend Allah's detailing of answers to questions of our origin. God informing of parking His throne upon water at the creation site is an emphasized detail. If only this detail of water at the creation site were to miss in the Quran, chances are that its narrative on the entire creation process would be undermined. How can this Quran be Muhammad's production? What does your highly read consciousness propose? Allah's throne **had been** upon water. **-Had been-** is a directive inferring displacement from the previously parked position which was near the solar system to elsewhere. After the creation of planet earth and its skies this position has shifted to His abode wherein gravity slows down time for a unit measure of a day with Allah equalled to one thousand years of our reckoning. [121] The extrapolation of Allah's throne parked upon water while creating earth and its skies tells of the significance of perfecting water bodies in the making of planet earth in our solar system. The naturalist's assertion that it must be a surprise; if we did not evolve here on planet earth in this universe because everything that went on to result us in the Cosmic arrangement just came in exact measures to breathe life and to create evolution on earth. Therefore, they conclude that "our being alive here is not at all a surprise." It is the most calamitous irony of knowledge failing to effect humility when it is most wanted! The Quran's tying of this instance in this verse *Quran 11:7 "...and His Throne had been upon water- that He might test you as to which of you is best in conduct..."* is telling of this teleology for assessing us which is among the evidence of Allah's achieving essentialities via a string of exclusive favours upon planet earth for resulting the exactness effected in nature to sprout life and for creating us for belief and conduct lest to face consequences is humbling.

> Quran 68:42-45 The Day when the shin shall be laid bare, and
> they shall be summoned to bow in adoration, but they shall not
> be able. Their eyes will be cast down, ignominy will cover them;

---

[121] Quran 22:47 ...verily a Day in the sight of your Lord is like a thousand years of your reckoning.

seeing that they had been summoned aforetime to bow in adoration, while they were whole (and had refused). Then leave Me alone with such as reject this Message: by degrees shall We punish them from directions they perceive not. A (long) respite will I grant them: truly powerful is My Plan.

Famous naturalist and Atheist Richard Dawkins would want to bow down in adoration to Almighty if the Lord showed up in the horizon as he is telecasted speaking in many of his popular interviews for he has come to opinion that such a being deserving his submission perhaps is only abstract. The reasons for misbelief are many as summed in this work. Perchance may be more, but his personals are a pejorative upon me to enquire; besides it is not up to guilty men to judge each other for matters of faith but God the absolute will reckon our primacies that deluded us from seeking Him. The day when Allah will finally show up, that day, the stiffened backs wouldn't be able to bow even if they profess to become ardent worshippers. What a bad trade! You talk the exact business (the miracle of creation) and yet are prevented to conduce a deal (with the Creator). And this book is an answer to Richard Dawkins's query concerning the seven skies mentioned in the Quran. The symbolic shin of Allah laid bare is an eventual fact communicated to unfold. In eastern cultures, the right shin is alluded to express clear authority. Such is the symbolism Allah uses to drive the point down the fallen awareness of man. For Allah's self is unlike human eye has ever seen.

Allah's Al-Arsh (the throne) is made of materials that scatter light in the visible spectrum[122] and the Angels on doomsday bear it in the near earth's celestial space. The kursi (chair) is situated above Al-Arsh (the throne) and is therefore understood as Allah's central pew of stately disposition over His Al-Arsh (the throne).[123] And further His Al-Arsh position moves about parking in the deemed

---

[122] Quran 69:16-17 And the sky will split, for that Day it is infirm. And the Angels are at its edges. And eight will bear the Throne of your Lord above them, that Day.
[123] The understanding that Allah's kursi (chair) is position over the throne (Al-Arsh) is also understood from: Quran 12:100 And he (Joseph) raised his parents high on the throne (of dignity), and they fell in prostration (exalting Allah), before him. He said: "O my father! this is the fulfilment of my vision of old! Allah has made it come true! He was indeed good to me when He took me out of prison and brought you (all here)

celestial space in the universe, as and when, needed to be acutely involved in the making of His creations and also for reckoning as told in verses 68:42-45 cited before. However, in Muslims traditions Al-Arsh's (the throne) superiority over the kursi (chair) is talked of as being like the given formula of the superiority of the desert over the finger ring. But as highlighted earlier in this chapter comparisons to Al-Arsh and kursi with any of the seven major Cosmic skies are not drawn in the Quran and further Al-Arsh is said to change positions within the realms of the universe. Thus, overstatements are noted to surround in proliferating the pattern formula given by Prophet Muhammad. Therefore, misperception surrounds in figuring right information entirely in this area from secondary Islamic sources. Viz. the Quranic impetus is clearly the dominant narrative. This concept is well understood and popular among learned Muslims to extract sufficing essentials from narrated content gathered in secondary sources. In other words, classifying content as authentic (Arabic: Sahih) from the voluminous hadith narrations is an ongoing effort over which the Quranic narrative is clearly sovereign.

---

out of the desert, (even) after Satan had sown enmity between me and my brothers. Verily my Lord understands best the mysteries of all that He plans to do, for verily He is full of knowledge and wisdom.

6.10 ALLAH'S CREATION ACT

> Quran 10:61 ...and nothing elusive is from your Lord, of the particle's weight within the earth or in the sky or (any miniscule) tinier than it or greater but that it is in a clear register.

As told in the verse, the summation of resulted creations in God's domain in the Cosmos are enumerated in clear records.[124] To emulate the command given by God for the creation of earth, earth's primeval matter (gas, dust & clay) undertake an effort after which several of those protoplanets take form. Then as the creation process is continued, the best is selected for perfection and others wither away from full realisations. However, it is my limited attempt here to explain things of which I barely have knowledge. It could be entirely different with God. Perhaps it could be said that they were created for wisdom & reflection like the bodies in our solar system for candidates such as Prophet Abraham, to make sense of the laws of universe: Venus, Mars and Jupiter could really have been planted. Or possibly for us today, to challenge our pride to comprehend knowledge of what fate befell on Mars and if God's-will mattered?

While we are surmising various theories to explain formations found in our solar system, our surface understanding of Cosmic sciences is teased by the dumbstruck display of dead planets and many variations in our neighbourhood. Initially, a small over-density in the gas cloud set the full cloud of nebula in rotation. Gradually the collapse gained momentum and gravitation took hold. This contraction under gravitation gave rise to hottest region at centre and birthed the protosun, with its total mass concentration of 99.8% of the entire solar system's mass. Sun had formed with its internal gas pressure exerting a force to balance the lead gravitational contraction. This and the nuclear energy expelled other elements and compounds outward clearing the spaces of heliosphere.[125] Remaining 0.2% of mass made of gas, dust & clay components of the nebula organising into many separate gravitational contrails along the disk of the protosun had begun the process of planet formation. Awhile when a shower of

---

[124] Also, 34:3 and 72:25-28.

[125] Title search: Lecture 13- The Nebular Theory of the origin of the Solar System on atropos.as.arizona.edu.

minerals & heavy elements coming from more massive stars were integrated into these clay formations.

In the science of fluid mechanics, disking of gases and dust into a net swirl about a massive gravitational centre such as of the protosun, should have entirely separated the lighter elements and compounds from the heavier such as hydrogen plus helium gases then with ices of ammonia plus methane & water and then with silicates; concentrating silicates closer to sun, ices outwards and the lighter gases still outwards. This explains the Gas giants—Jovian set of planets to have arranged further away from the sun while we (terrestrial planets) occupy near spaces made of dense silicates. But findings of Kuiper belt (also Pluto's residence) & Oort cloud after a series of gas giants does not accord well with our understanding of the behaviour of matter in protosun's disk. A clear conscious design of God is evident here in preparing the frontiers and mounting them with asteroids and other bodies for purposes of protection, reconnaissance and containment of jinn interference (discussed in chapter 7). Further this separation & concentration due to protosun's gravitation & intense net swirl would mean there would be one gigantic terrestrial planet made of silicates. And one low-density planet after the frost line, where due to low temperatures of fading sun's astrosphere, ices condense & here in this transit zone a disk of asteroids and comets can tuck in too-well. And finally, one humongous gas giant further away from the other two planets could explain the behaviour of matter given to intense swirls. As seen, it is clearly not the case with how matter has coalesced.

The theory that due to Jupiter's influence the asteroid belt did not form into a planet is not selling because of the timing argument as Jupiter too was "accreted" from zero. Then in this period of 50-100 million years per current science estimates, asteroids in the asteroid belt stood a chance to accrete as well as the argument of Jupiter's gravitational influence is clearly post Jupiter's formation. This also contests current hypothesis in place of planet formations via accretion of flakes by gentle collision to be an unyielding research view; given that other forces such as gravity contrails (mechanics of gravity) influenced formation of protoplanets like the gravity contrail leading to contractions of protosun. The many separate terrestrial planets with iron cores and Jovian planets postulated to have rock cores surrounded by liquid metallic hydrogen is suggestive of separate gravity contrails to have formed these protoplanets by the dominant forces of these contrails acting over and above the centrifugal & Coriolis forces in the plane of swirl encountered in the far spreading disk of protosun. Protoplanets were thus birthed. This therefore counters our understanding in current science of planetesimals forming by accretion in the sense of flakes to flakes collisions.

For an amateur like me who only serves curiosity; it is an uneasy area to argue several possibilities as experts in the field of astrophysics too fumble in postulating the origin sequences. We can however try to make sense of how what all exists has come to exist: *Quran 40:57 "Clearly the creation of the skies and earth is greater than the creation of mankind, but most of the people do not know."*

It can't be simpler told than this. In this end to end planning by God; thus, to orchestrate the size and content of the Universe as we come to terms with God's design. It is not mere play for Him but in fact it is clearly a serious affair! A lone chance awarded to us in the never endings of the Cosmos with the gift of intellect which could be held accountable for gross error is never failing of God's signatures for justice. Therefore, it is for us to contemplate: Is it worth to consume our lives in play by deserting this Creator of ours?

> Quran 27:64 Is He (not best) who begins creation and then repeats it and who provides for you from the skies and earth? Is there another power besides Allah? Say, "Produce your proof, if you should be truthful."
>
> 65. Say, "None in the skies and earth knows the unseen except Allah, and they do not perceive when they will be resurrected."
>
> 66. Rather, their knowledge is arrested concerning the afterlife (in Paradise planet). Rather, they are in doubt about it. Rather, they are concerning it, blind.

Can this impeccable scriptural confidence surmise that we discover the locus of Paradise planet as we learn by our sharp lenses to peer into the Cosmos? Of the creation processes supernova explosion is counted as one of the ways for newer formations of protoplanets or even for providing essential elements & minerals for distant located protoplanets within the realms of those respective first Cosmic skies. Citing supernova explosion of the crab nebula in Perseus arm which could make up for the dressings of rich materials needed for the emergent Paradise planet in the interiors of that starry colony by an analogy of supernovae explosions providing for earth's dressings here in Orion arm. Thus, arguing the locus of Paradise planet to be found in our neighbourhood Perseus arm of Milky

Way. It perhaps may be our destination where we will rejoice or suffer (see **Figure 22**). However, in absence of clear evidence, this work does not propose Paradise planet to be located in Perseus arm of Milky Way but explores it as a possibility among possibilities.

God's commands result in the making of similarities across the universe and then His acute-self is involved in perfecting targeted manifestations such as the planet earth and its skies for a clear definite event. Per His intricate laws of creation, His orders for perfecting a given specific, constitute factors required in perfecting it and for proceeding in its making. Hence accomplishing the required exactness by His might, the specific, like the earth, is brought to realization via a string of exclusive favours graced upon it from Him. As highlighted in previous section Allah's Al-Arsh (the throne) was positioned over the water in the near celestial space during the creation of planet earth and its skies. There it was God's most likely previously parked operational locus as postulated in this work to be in our upper deck area in the Orion nebula of the Orion constellation, a subunit within the spiral arm of Orion. This is when Allah had meticulously designed our planet earth in the solar system. Then to repeat creations such as to create the Paradise planet say for example in Perseus arm of Milky way at a distance of 6500 ly plus from earth; His signatures are automated wherein His personal descent is not deemed necessary: *Quran 30:27 "And it is He who begins creation; then He repeats it, and that is (even) **easier for Him**. To Him belongs the highest attribute in the skies and earth. And He is the Exalted in Might, the Wise."* Though this verse is specific in talking of creation of beings the law holds good for all types of creations in the Cosmos.

As detailed God's parking at the creation site of our solar system in the Orion nebula was participated by legions of Angels spearheaded by Gabriel who were stationed with God in this horizon for the first time as God had made an approach with His entourage known as 'الملإ الأعلى (al-mala-il-āla)' meaning the high assembly. And then long after this event at the fore of human activity on earth, Gabriel descended to Sirius for duties of divine communication from his held stage in Orion. How better than what is said: A wholesome narrative of skies can be found in a scripture of the old? Even as your hearts leap for the Quran, perchance the White tribes may come to bear it lest for race, language or utter neglect.

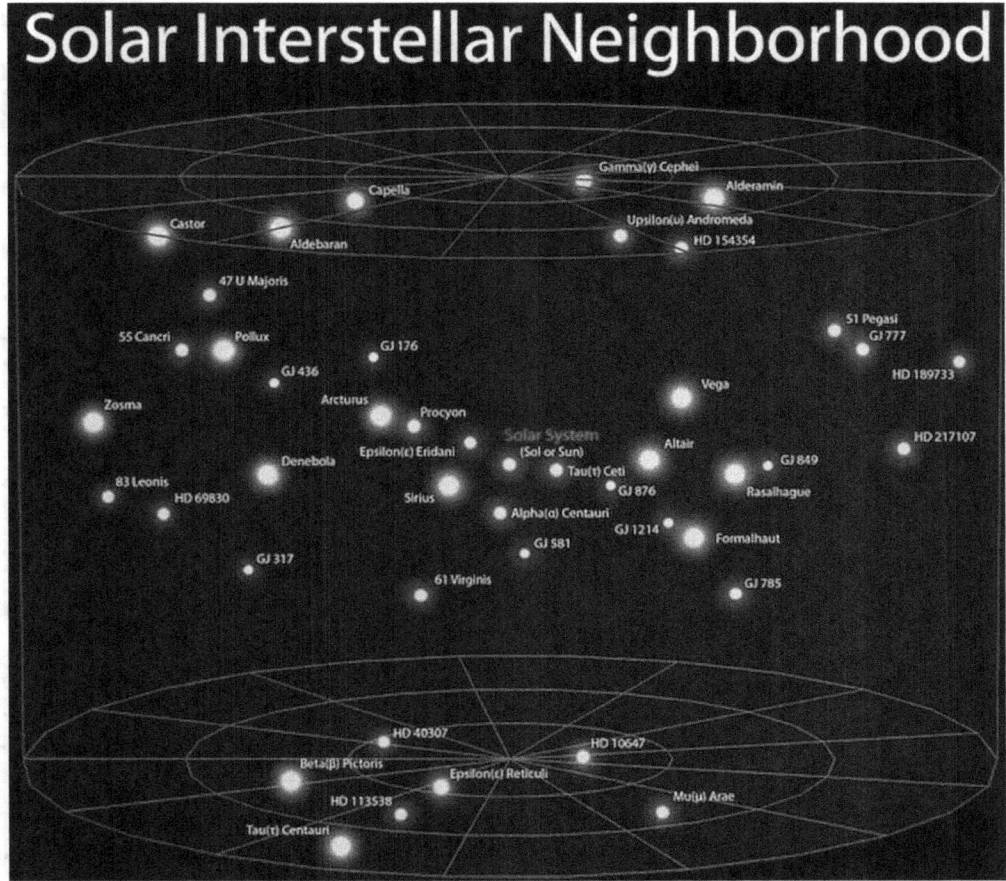

Figure 15 *Solar Neighbourhood- Only a tiny volume of 15 ly.*

*Few prominent stars of solar neighbourhood within 15 ly, in the first Cosmic sky, solar-system at the centre. This idea of solar neighbourhood depicted here does great injustice to the actual composition of solar neighbourhood talked in Quran and observed in the celestial space defining its boundaries in the Orion spur to at least 3500 ly in volume, see **Figure 10**, **Figure 30**, **Figure 31**, and **Figure 32** for a real feeling; Image sourced: Wikimedia Commons; Credits: Andrew Z. Colvin.*

Figure 16 *Milky Way- Locating solar neighbourhood.*
Diagram of the Milky Way a spiral galaxy, with a pointer showing sun's location; Image sourced: Wikimedia Commons; Credits: Andrew Z. Colvin.

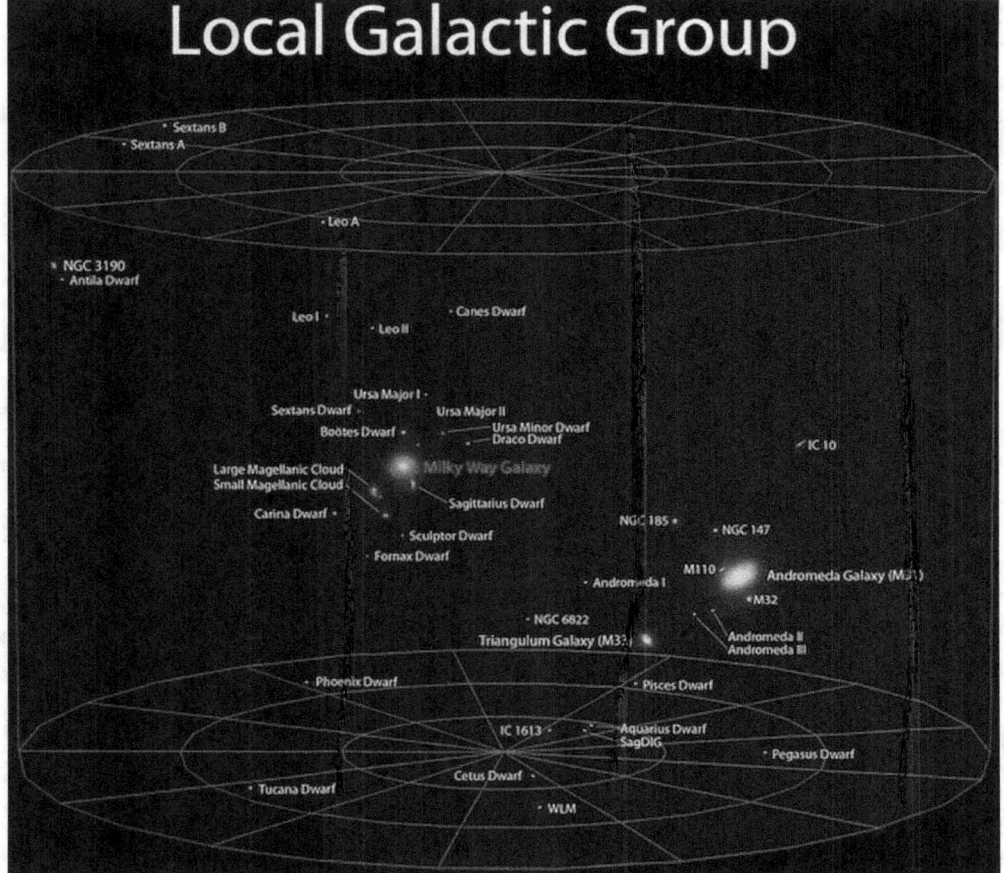

Figure 17 *Local Galactic Group.*
Diagram of the galaxies in the Local Group relative to the Milky Way and Andromeda galaxy shown as tiny finger rings lost in a desert; Image sourced: Wikimedia Commons; Credits: Andrew Z. Colvin.

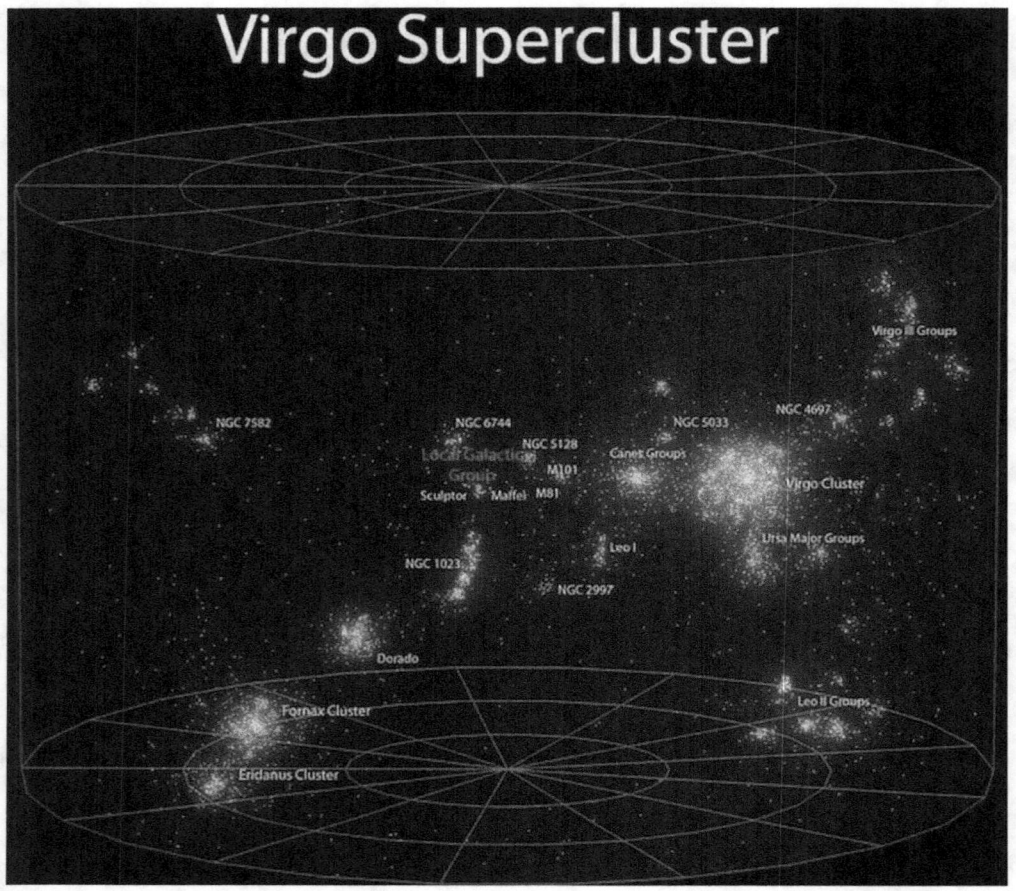

Figure 18 *Virgo Supercluster.*

Diagram1- of the Local Galactic Groups in Virgo Supercluster; Local Group containing Milky Way relative to Virgo supercluster is like a tiny ring lost in the vast desert; Image sourced: Wikimedia Commons; Credits: Andrew Z. Colvin.

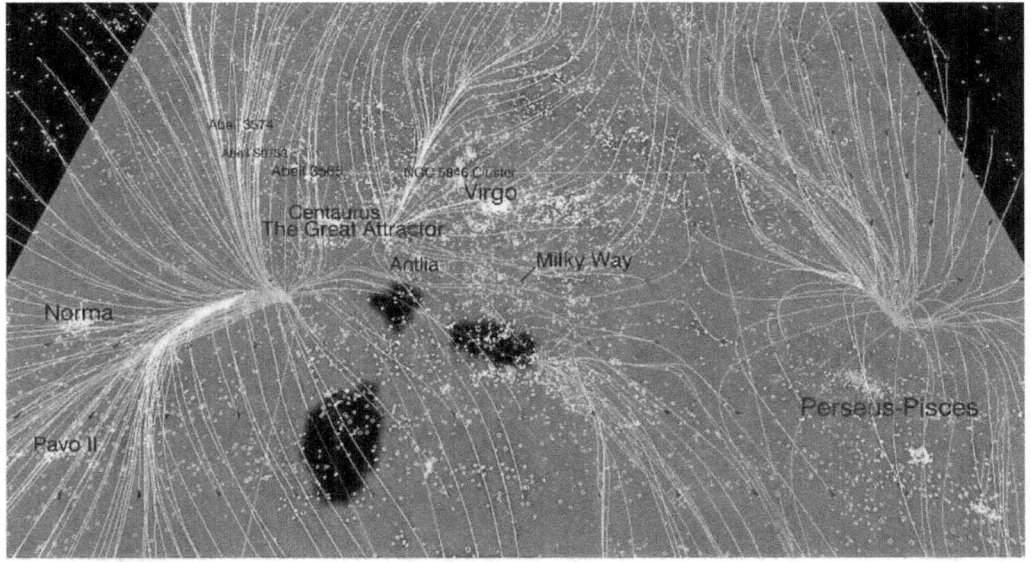

Figure 19 *Laniakea Supercluster- The Immeasurable Heaven.*

Diagram2- Laniakea Supercluster; the greatest overdensities (roots of tendril concentrations) and underdensities (in black) are gravitational differences. Image sourced: forbes.com. Credits: National Radio Astronomy Observatory (NRAO); public.nrco.edu. Also, SDvision interactive visualization software by DP at CEA/Saclay, France.

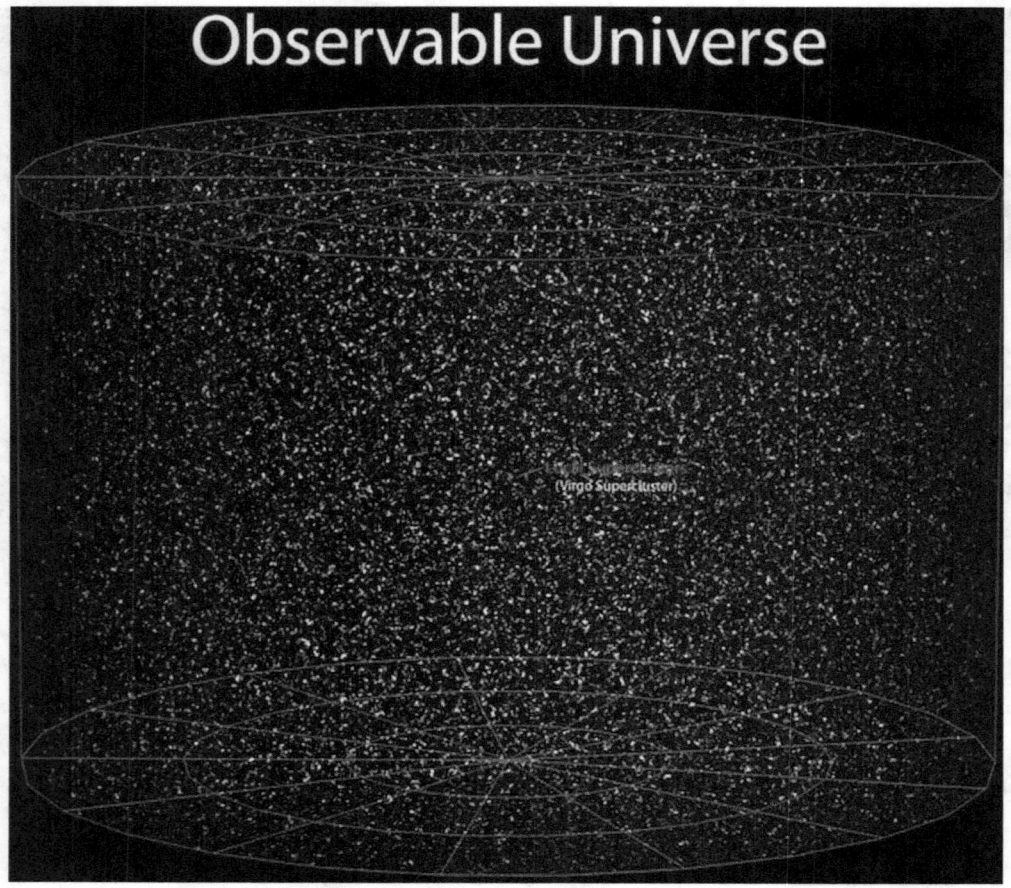

Figure 20 *Observable Universe- Needle worked with strings of rings.*
Diagram of the observable universe relative to the Local Supercluster (Virgo Supercluster);
again, Prophet Muhammad's ring and the desert formula is apt; Image sourced: Wikimedia
Commons; Credits: Andrew Z. Colvin.

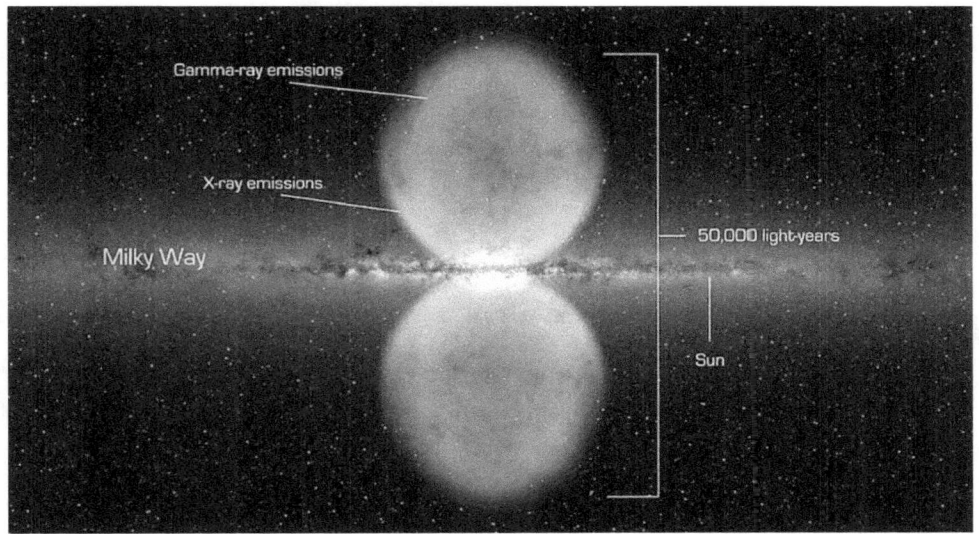

Figure 21 *Gamma-ray aura at galactic centre.*
Gamma-Ray bubble at the centre of the Milky Way; Credit: NASA's Goddard Space Flight Centre; Image sourced: Wikimedia Commons.

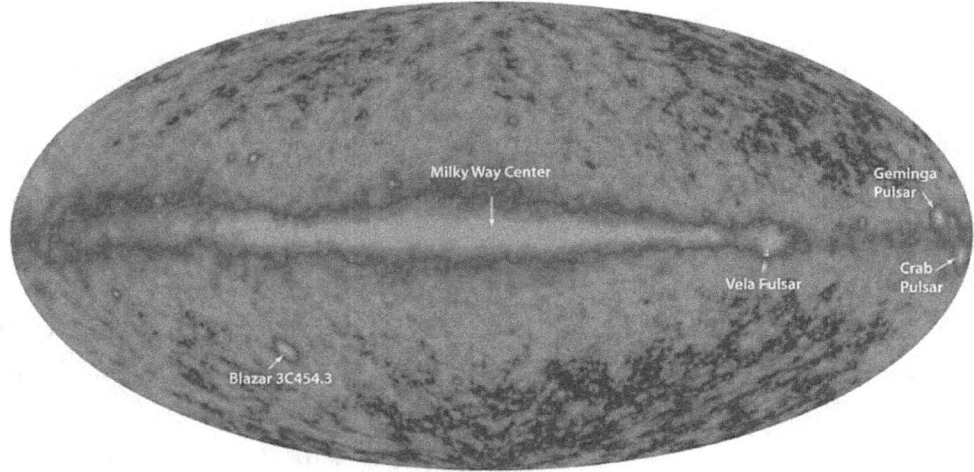

Figure 22 *Position of Vela and Crab pulsar in Milky Way.*
Vela pulsar in Orion arm is located 1000 ly from Sun within First Cosmic Sky. Crab pulsar does not belong to Orion arm, is located 6500 ly away from Sun in Perseus arm; Credit: NASA's /DOE/International LAT Team; Image sourced: Wikimedia Commons.

7 OUR WINDOW INTO THE UNIVERSE

*Window spaces help breathe a sense of spirit in lively bodies; living
beyond death is delusion for the naturalist. Alas! For them Cosmos
from far is thundering...*

When travelling through the pathways of the Universe, Angels find destinations by stars and celestial constructions. Like us on earth, they too are guided by stars towards their directed terminuses. The sixth Cosmic sky i.e. the sky possessing webbed pathways is at the heart of the Cosmic travel maps (section 6.8). In these highways, Angels route their travels across filaments, clusters of galaxies, constellations and stars in deep space making prominent veering signposts proclaiming this massive artistry in creation is a serious business! Condemning men and jinn to eternal punishments for disbelief isn't mere play for the Almighty Cherisher who has lovingly given us life. Dearly loved-life is His exclusive grant. Testing faith from the recipients of freewill is therefore due on Him and nothing but arrogance bears the blameworthiness of rebellion to Allah. Liberty to choose or refuse is clearly in for justification and is a natural outcome to freewill. If we honour this rationale, Allah's existence is manifested. For untaught who wouldn't avail to argue for God, this is enough for siding belief in the creator of Cosmos, the giver of life who in position is located far from us: "...*Had We (Allah) so intended to take a pastime, We could have taken* it from (what is near) with Us- if (indeed) We were to do so..."[126], but in reality is also entangled to us nearer than our jugular vein: *Quran 50:16 "...for We (Allah) are nearer to him than (his) jugular vein."* This entanglement is by the life-line substance الروح (ar-ruh) meaning the spirit, a real elusive thing: *Quran 17:85 "And they ask you (O Muhammad), about the spirit (also written as soul). Say, "...mankind has not been given of it much knowledge."*[127] It is told in the Quran, if this spirit or 'soul' when retrieved from the material organisms then demise prevails. Owing to our inability to disprove Allah's plausibility, disbelief therefore is condemned to eternal consequences. Justification is the only substantive for exercising freedom and liberty, discounting the need to justify warps the boundaries between truth and error. There couldn't be a better test-bed than what is found here with us to evaluate faith in the Unseen Lord- who is from far compelling yet palpably perceivable. But for the blind the Cosmos is perhaps only the tangible material existences; although in modern awareness much is now known to remain at large from being detectable irking the mesmerised naturalist apologists.

> Quran 52:35-38 Weren't they created from nothing or were they the Creators? Or did they create the skies and the earth? Rather, they are not certain (of belief).
>
> Or do they possess depositories of your Lord? Or are they the controllers (of resources)? Or have they mechanisms (to pry) upon which they listen? Then let their listener produce a clear authority.

---

[126] Quran 21:16-19 And We (Allah) did not create the skies (earthly) and earth and all between them in play. Had We (Allah) so intended to take a pastime, We could have taken it from (what is near) with Us - if (indeed) We were to do so. Rather, We dash the truth upon falsehood, and it destroys it, and thereupon it vanishes. And for you is destruction for what you describe (wrongly) of Us. To Him belongs whoever is in the skies and the earth. And those near Him are not prevented by arrogance from His worship, nor do they tire.

[127] If the Arabic word 'ruh' is translatable for 'spirit' and the word 'nafs' for 'the self' or 'soul', then the 'evil spirit' as a phrase in English is annulled. Because 'ruh' or 'spirit' in the Quran is only —the life-giving unperceivable element while it has nothing to do with good or evil. Whereas 'nafs' the self or also written as 'soul' at times for language purposes can be and is most fittingly good and evil.

In the ways (sunan or sunnah) of Allah His laws makeup the fundamental nature even of those which are not apparent to us or are supernatural or spiritual. In defining "supernatural" if there is a concession to reduce its substantive from reality then I would prefer to use the term -spiritual- and define them as "entities which interface with recipients of freewill by intricate ways of direction from God" than to fume ghostly or spooky phenomenon. God has enumerated His sunan (many governing phenomena) in clear words in the Quran. It is God's Will which eventually comes to prevail. Perhaps due to ignoring observations of the deeper ways of how nature organises; as a result, many grow adamant and believe that spiritual sources do not contribute to **evidential nature.** Hopefully as we progress to comprehend more to model our lives better, we'll inch closer to learning the natural phenomena driving in from the spiritual dimension. There are however limits and some matters reside outside our domain (such as the elusive dimension of dark matter) and maybe it will remain unseen & unfathomable and thus would not deduce naturally to us. In God's law of guidance to mankind, He awards a nation by favouring them over others in turns, gives them a revelation and supports them making an example to the rest, and then he subjects them to trials to test who among them witnesses the truth, guards-faith, conducts righteously and remains thankful. In one such trial He sent a couple of Angels to try the people in Solomon's kingdom by teaching them magic only to barter in exchange for their faith. Those who traded their faith for Magic and pursued secret arts, popularly known as Kabbala chose to rebel against Allah for some material gain. In freewill its recipients can with choice worship Allah in servitude or reject Him. So did Satan rebel against Him for his jealousy for Allah's bestowal of honour to mankind. And so, do many among men in pursuits of self-appraisal disregard Allah and despite largely failing to produce a credible argument to refute the plausibility of His existence they make a free choice to disbelieve.

7.1 MAGICAL SKY- OF MESMERISING ARTISTRY

The Quran, because of its mind-blowing prose in classical Arabic demands a critical study. Most possibly, clerics and critics alike will be at fault miserably if they rush to interpret it lacking reflection. This is the reason why most often you'll find people jumbling the Quranic verses or selectively singling out narratives without proper context owing to their substantive illiteracy. They're also found to ignorantly flout contexts and mix-up cross references. As a result, they find Quranic narrative to be discontinuous. Contrary to their desires to label the Quran the way they do, the Quran is clear and elegant even to non-Arabic subscribers. Besides diligence, impulse to inquire and an ardent desire to learn, it takes humility more than any other criteria to realise the truth of the Quran. If God sees your worth, He will get you to drink from an ocean of knowledge and show you a glimpse of the greater reality. Those who drink from the Quran to them the drivels of ignorance become obvious, their arguments are endued in power and their contenders fade away. They grow up to familiarize with truth as if to adapt and they will unstring from hearsay, subduing bias and prejudices with ease. They wouldn't go in circles trying to assert or be in denial for matters which have already been made clear say a thousand years ago. They would critique approximate evidences and superstition. They won't see the magic of random stimulus but instead will submit to God the source of All Knowledge and Wisdom.

> Quran 15:10-11 And We (Allah) have certainly sent (messengers) before you (O Muhammad), among the sects of the former peoples. And no messenger would come to them except that they ridiculed him.
>
> 12-13. Thus, do We insert (denial) into the hearts of the criminals. They will not believe in it, while there has already occurred the precedent of the former peoples.
>
> 14-15. Albeit if We opened for them doors of the heaven and they continued to ascend therein. They are sure to say, "Our eyes have only been dazzled. Rather, we are a people (in exploration) fascinated."
>
> 16-18. And We have placed in the sky starry constructed mansions and have beautified it for the observers. And We have protected it from every devil expelled (from the mercy of

Allah). Except one who steals a hearing and is pursued by a clear meteor (burning flame).

The delicate balance inherent in every conscious mind is quantified for a favourable response towards God. When a person drops conscience and detains truth Allah's guidance is then tactically severed. Unless the target leaps back penitent and then commits to amends observing facts more than fiction; there is hope. If material existences become the only pursuit of their mind, then Allah seeps in denial of truth into such minds. ***"Thus, do We insert (denial) into the hearts of the criminals."*** Literalists must know that it is not the muscular heart alluded to here in verse 15:12 but an idiom is pointed to the neurons in the brain allocated to functioning of the hearts of men and women. Thus, blockades at the level of neurons are set in to impede crucial cognitive dissonances upon a person's denial of truth. Truth is a construct built on facts and of rationality in the mind. Thus, when dearly held blind-beliefs are confronted with truth, bravery and courage are used to demonstrate the calibre of one's heart in its fight against bias. If any poorly shows prejudice, it is clearly a bad start. It is not the muscle itself (the heart) which when needed is operated for replacement, but it is the mind: *Quran 75:2 "And I swear by the reproaching mind."* This mind which **upbraids** if put to use it makes and defines the brave-hearts and the true-will. Generations of humans have already succumbed to desires and Allah did do-away with them. If we imitate in similar footsteps a similar end is inevitable.

Verse 15:14 *"Albeit if We opened for them doors of the sky and they continued to ascend therein"* cited herein technically opens a futuristic encounter. Humanity in seventh century did wonder on the beauty shown in the sky but, opening doors of the sky was atypical paradigm for them. Allah the foreseeing tells: "Even when, If I open for them doors and windows to gaze through the Cosmic skies and permit them to ascend therein, they would still mostly disbelieve." It is an amazing detail. Hardly any man perhaps could have made such a statement fourteen-centuries ago. It is not the sight mesmerising artistry and display of material that helps conceive belief. But at a fundamental level it is only the remembrance of the Holy name of God, the grantor of life. Whom if we dismiss for pride, no amount of evidence then will make for a convincing aid. Even if the dead were to rise again and speak, disbelievers will still be short from belief: *Quran 6:111 "Even if We (Allah) did send unto them Angels, and the dead did speak unto them, and We gathered together all things before their very eyes, they are not the ones to believe, unless it is in Allah's plan (to compel belief unto them). But most of them ignore (the truth)."*[128] It is the rationality of the mind that conceives to concede to faith. Denying merely for the sake of denying will harden the neurons, once hardwired this pattern in the brain then averts all attempts to recognise evidence and no amount of truth becomes compelling enough. Even the blind of an eye believes! As rightly said in the Quran, even miracles aren't a premise for belief, but it is the use of mind in a wise way which helps to testify to the truth of God's existence. For the materialistic kind who do not want to see beyond what their physical eyes see, it is impractical a request and less benign for Almighty to be anchored up the sky to be seen for all times. However, the law invoked by God to assess belief from the recipients of freewill works as it is in the unfolding now. On account of inventions of telescopes and space access the doors and windows into the Cosmos were struck open. Humanity was prided as we ascended into space and were mesmerised by the display and sheer magnitude beholding the Universe. Telling of facts many observers despite their spell bounding experience haven't believed. Prejudice and bias in circles of peers adds to delusions and refrains many from sincere acknowledgements of Almighty. There is a sense of blindness and false hope for their forestalling belief until for their discovery like something from an Almighty or for Himself to manifest so that they could unfailingly believe. Be that as it may, Almighty God has remained unseen in past and is likely to remain so, until this grand test casted for humanity and jinns expedites.

Kabbala is yet another extreme. Squarely opposite to Atheism. The science of Kabbala upsets belief by the choice of dissolving it, having men traded numerous gains of secret arts in exchange for faith. Thereafter their interests are shifted to anticipate Allah's demonstration of retribution to mankind for disbelief and ill-conduct. And, for their imminent grief in the event Allah provides guidance to people as and when people would fight bias to concede servitude before Him. Thus, God alone bears the prestigious entitlement, for the opportunity of precious life and grant of guidance. Kabbalist's come to believe to live and enjoy a life free from servitude to Allah. And recline to enjoy black magic and secret arts leading to the pursuit of elusive sciences.

---

[128] And, see Quran 13:31.

It may not be an overwhelming reality today, as our knowledge still hasn't sought verifiable techniques to interface with jinns. At this stage, it works via trust and belief in the practice area of secret arts. But Allah has indicated the full extent plausible for interface between men and jinns for confederation, as shown below:

> Quran 15:88 Say, if mankind and the jinn gathered to produce the like of this Quran, they could not produce the like of it, even if they mutually backup support.

As we develop technology for Alien Communication, it may perhaps bring us nigh to the rebel jinn's dismay. They may feel compelled to reveal themselves and thus be subservient to us like with Prophet Solomon (see section 7.4), akin to every other creature on earth under the firm grips of mankind. Then that collaboration could lead us to establish a proper interface for joint endeavours even if it be in rebellion for prying and discoveries of -safer- techniques from the jinns knowhow to tap classified Cosmic wires beamed from conferences of the high-assembly down to the nether world. Joint or a lone hunt if but launched by men is called into questioning by Allah in a verse: *Quran 52:38 "Or have they a stairway (ability to launch into skies and pry), upon which they listen? Then let a listener of theirs produce a manifest proof."* Elucidating space ventures, Allah simply poses, if your claim of tapping My classified information is true then produce a clear proof. This shows Allah's terrific artistry in the Cosmic skies and His protective gears that are working all day long guarding the networks of information flow. Along with stars with high intensity discharges of enormous radiation and deadly bursts of energy there are incinerating cinders that are laced into chases all of which are necessary for the architecture of Cosmos to blockade leaks in incoming communications. The information that in a galaxy apart from regions of sparsely populated stars, solar system is situated in a zone which is teaming with stars is an amazing finding told in the Quran indicative of verses 37:6 and 67:5, *"Allah has certainly beautified the **sky of this world** with glowing lamps (bright stars) and space objects in whorls"*, see **Figure 9**, **Figure 10**, **Figure 30**, **Figure 31**, and, **Figure 32**.

The deployment of prying infrastructure in this zone by the jinns to manipulate human affairs required blockades and other stringent measures be placed. Postulating rich star density and the buzz it generates perhaps in further upper limits of the EM spectrum (longer radio wavelengths) to barricade a specific band of incoming waves analogous to a jammer of incoming signals and to quarantine jinn space probes for any healthy reception is baselined by the evidential design found in this patch of the sky. High energy & radiation emitting pulsars, binaries or multiple star systems, star implosions and hammer heads of extreme brightness like Sirius with a companion battering make a collective and effective pack for protection. All-star astrospheres inherently are effective road blockers and appear to continually ward of jinn's persistent snooping efforts to approach a reception at the star-posts. These measures within the first Cosmic sky were effective to truncate jinn interference into humanity besides several other necessities as discussed earlier to secure the working of human affairs as told in the final revelation of Allah.

Jinns are deluded by their peer Satan who owing to his prolonged life (till the day of Judgement, Quran 17:61-65) exercises enormous power among his crowds. To the Jinns, Angels are perhaps perceived as another alien species in horizons of space akin to the widely found humans on earth. And therefore, building this narrative, Satan accuses Angels as murderers of Jinns and the many species on earth. Deluded by this and perhaps hopeful of attaining the seemingly immortal trait of their leader, even non-partisans among jinnis are thusly misled from worship of Allah. Being a far advanced species than us, their reasons for believing in the implausibility of the Almighty are at least as creative (if not more) as ours. Thence, cynicism deepens, and belief becomes abstract before material existences, all these advances to doubtfulness, randomness and disbelief.

Scientific community has its own historical cysts. Piltdown-man show at the turn of the 20th century swelled the high-tech print media with Darwin's delusions of survival of the fittest and natural selection contradicting the determination of God for creation of species in their selected natural habitats. The freeing impetus of this lie has forged modern minds to rebel the Most Gracious (God) and thus are deluded to uncritically lean towards random forces at work. Atheistic young minds graduating from secondary schools undertake research in their already narrowed thought line and therefore see evidences from an inherent bias. Likewise, discriminatory are delusional sects making schisms in religion by the straitjacket attitude valuing closed circle opinions greater than impending justified outcomes from intellectual public discourses. If not for this whim of disbelievers in Christ, followers of Christ would not be known differently. An indivisible part of Jewish culture and traditions had to part away for a new title by bystanders; people from Antioch named them first as -Christians-.[129] Rather Jesus with his Jewish fellowship would only be known by Jewish heritages. Jewish rabbis differing over Jesus for possessions of stage and glory struck a barrier and built a new religion. Transparency is the gift entrusted by Allah upon mankind demonstrated by His Prophets calling without discrimination to all seekers of truth as can be seen in revelations of the Quran inviting all children of Adam. In our present world, it is exceedingly clear that this force of discrimination has divided Almighty God into a sectarian-ally.

---

[129] Book of Acts 11:26 ... and it was in Antioch that the disciples were first called 'Christians'.

7.2 METEORS ASTEROIDS AND COMETS

Herein are discussed Meteors, Asteroids and Comets in the light of the Quran. In celestial mechanics, there are five positions in any given orbital configuration between two large bodies known as L-points or Lagrange points, see **Figure 23**. Non-planetary debris affected only by gravity can maintain a fixed position relative to the two large bodies in these five spatial grid locations. It is cited in the Quran that Jinns conversed regarding their losses in real-estate in these stable spatial grids from where they used to tap into communications. These now stand occupied by stern guards and space objects: *Quran 72:8 "And we (few fellow jinnis) had (recently) **accessed the sky** but found it filled with stern guards and meteors."* Remarking that if any now try to listen, then such will be ousted from these stations: *Quran 72:9 "And that we used to **sit therein in stations** for listening. But whoever listens now finds a meteor lying in wait for him."* This information was brought to knowledge of men, in the beginning of seventh century during the Makkah phase of the Quran's revelation, a statement from the creation saga to those who are still awaiting communication from the Creator!

*Excerpts from Wikipedia: Several planets have natural satellites (space objects) near their L4 and L5 points known as trojans with respect to the Sun. Jupiter, particularly has more than a million of these. Artificial satellites have been placed at L1 and L2 with respect to the sun and earth, and earth and the moon, for various purposes. And the Lagrangian points have been proposed for a variety of future uses in space exploration. It is common to find objects at or orbiting the L4 and L5 points of natural orbital systems. These are commonly called "trojans"; in the 20th century, asteroids discovered orbiting at the Sun–Jupiter L4 and L5 points were named after characters from Homer's Iliad. Asteroids at the L4 point, which leads Jupiter, are referred to as the "Greek camp", whereas those at the L5 point are referred to as the "Trojan camp", see **Figure 24**. The Sun–Earth L4 and L5 points contain interplanetary dust and at least one asteroid.*

Most asteroids are between Mars and Jupiter in the asteroid belt whose make is thought to contain metals. Comet's make is mostly of flammable substances and of ice loads these are understood to come from Kuiper belt and Oort cloud, see **Figure 25**. Kuiper belt surrounds the exteriors in immediate vicinity of solar system. Whereas Oort cloud is a thick bubble of icy debris surrounding our solar system spreading outwards with its inner ring located at 2000 astronomical units

(AU),[130] this distant cloud may extend up to 100,000 AU which is almost a quarter of the way to the nearest star, Proxima Centauri. It is spherically shaped and consists of an outer cloud and a torus (doughnut-shaped) inner cloud, see **Figure 25** and **Figure 27**.

There are many theories wrongfully attributed to be the cause for dislodging comets from their orbits from within the Oort Cloud. Thinking that the gravitational pull exerted by the passage of stars, nebula or by events in the Milky Way have an affect amounting to their total dislodgement into a free fall towards earth or elsewhere is simply naive. These actions if influential must have necessarily knocked out large proportions of objects out of their orbits due to their spheres of influence sending them on a headlong rush toward the sun or out into the void of space which has not happened. Our solar system is still dressed in these ornaments in its exteriors and amidst its planetary orbits. So, what then could be the cause? The creator of Cosmos reveals the true reason in the Quran which is a step closer in discovering the actual physical cause. Comets, asteroids and meteors are ejected on precisely defined missions. Whether they be from the exterior boundaries of solar influence like Oort cloud or Kuiper belt or from within asteroid belt and the Lagrangian locations within solar influences, these objects are flung to surmount chases of intruding entities towards near earth or near solar orbits by stern guards. These are interplanetary missiles meant to mount a castigatory offensive on the eavesdroppers. Such is the actuality of the matter revealed in the Quran as one of the many reasons for the phenomenon which mainstream scientific community has adopted erroneously as having been caused by gravitational perturbations.

As we are mostly prone to learning the hard way, swimming across waves and waves of errors gradually crawling toward truth, our little understanding is subject to change after painstakingly being late. Thus far in the light of the Quran شواظ (shuwaz) is best interpreted as **space objects** in a trajected reconnaissance with a potential to fuel long missions loaded with burnable substances: *Quran 55:33 "O communities of jinn and men- If you can pass to the boundary regions of the skies and of earth, then pass. You will not so pass except by permission (of*

---

[130] One AU being the distance of earth from the sun roughly 93 million miles/150 million kms.

Allah)." Asteroids and Comets therefore with packets of ice loads, dust, rocky materials and organic compounds like methane and ammonia descriptive of *Quran 55:35 "Upon you be sent space objects of a fire brand and of smoke. And you won't be defended"* fall under the category of شواظ (shuwaz). From there on, determinisms of a chase or destruction is evoked against the transgressing jinnis and their elusive infrastructural developments. When these space objects approach proximal zones in star astrospheres, they are set alight and we perceive them as active asteroids or comets with huge tails. These eventually exhume after many such fierce missions. On contrary, the term شهاب (shihaab) is best interpreted for earth bound **space objects** characteristic of a stealth phenomenon that aren't on reconnaissance but are laced spontaneously to prohibit interferences instead of defined trajectories: *Quran 72:9 "And that we used to sit therein in stations for listening. But whoever listens now finds a meteor **lying in wait** for him."* Therefore, they aren't as massive as large asteroids or comets are and usually burn up when they make contact with earth's atmosphere. Meteors surmounted by swift travel speeds for ambush are set after jinns. Both شواظ (shuwaz) and شهاب (shihaab) do get sent to intercept jinns and tellingly upon humans too if we are targeted as trespassers.[131] Be as it may, our affairs in space are currently limited to gazing at the wonders of Allah's artistry for which the doors are open and permissions granted.[132] Though we are listening to sounds of the Cosmos chirping to us via our radio satellite stations, we aren't yet aware of tuning into classified wires of which jinns are already in pursuit and exploration.

Critics often point towards informal Islamic sources to force interpretation on the Quran. Wherever Hadiths are found describing over or under-stated scientific phenomenon they in the footsteps of uncritical Muslims immediately stoop low to authenticate those narrations to infer the Quran's incapacity to describe modern discoveries. Conscious Muslim believers (Mu'mins) aware of Allah's narrative and wisdom, haven't yet given up scrutinising the narrations alleged in the name of the Prophet which make up the volume of Hadiths. Reflection is the foremost examiner and the Quran is the litmus to authenticate information from

---

[131] Quran 55:35 uses a dual form of pronoun addressing both jinn and men communities.
[132] Quran 15:14 Albeit if We opened for them doors of the vaults (skies) and they continued to ascend therein.

Hadiths. Calls to consider Hadiths authentic (sahih) equally on par with the tone of the Quran's prose is an exaggeration and a grave sin. Even the earliest and most intimate companions of Prophet Muhammad would always add a disclaimer while quoting the Prophet. They clearly believed whatever they narrated quoting the Prophet was in their respective use of vocabulary and therefore were words not in par with the Quranic recital or even the words of the Prophet to the letter. Their narrating from the Prophet were not words that needed to be taken verbatim by later followers. Nothing but only the Quran is an error free file. It is undoubtedly an element first commissioned for belief. The Quran is the only scripture constantly under the guardianship of Allah preserved from adulterations. Prophet Muhammad has been quoted for many fabrications by a range of people. Among his nearest companions Umar Al Khattab, the second Khalifa (English: Caliph) continually cautioned new comers to faith to abstain from spurious quoting of the Prophet. Suppose if Muslims today are concreting the view that "sahih hadiths" are verbatim true and on par with the Quranic text then they are blameable to have ascribed an attribute of which Allah deemed it appropriate only for His divinely revealed book, whose authentic nature still is under His guardianship. Thus said, there are many agreeable popular Prophetic narrations (Arabic: Mutawatir Hadiths) needing no book references, if they can make sense of the historical reality and not contradict the Quranic wisdom then truth can be deciphered from it. From such widely confirmed narrations among peoples of faith correlating well with the Quran, information is relayed of flames or missiles chasing jinns which they saw shooting across the sky. One of these Mutawatir hadiths, clarify that even earliest Muslims were clear that meteors and stars were different objects unlike what critics usually allege. The narration goes "…then the dwellers of heaven seek information from them until this information reaches the heaven of this world. In this process of transmission (the jinn snatches) what he manages to overhear, and he carries it to his mates. And while the Angels see the jinns, they attack them with meteors. If they narrate only which they manage to snatch that is correct, but they alloy it with lies and make additions to it."[133]

---

[133] This is sourced to Abdullah. Ibn 'Abbas a nephew of the Prophet, in the Book of Muslim ibn Al-Hajjaj 26:5538.

Contrary to the allegations of its critics the Quran clearly differentiates between stars and meteors. The Quran uses three different words to allude to stars: النجم (an-najam) a proper noun for stars and then كوكب (kawkab) and مصباح (misbah). The كوكب (kawkab) can also be used to mean planets and space objects as discussed earlier in chapter 3. But مصباح (misbah) has only one technical connotation which is that of a glow of a lamp of light. Therefore, it is always translated as glow lamps. But the word شهاب (shihaab) translated as burning flame in many Quran translations is clearly a meteor. شهاب (shihaab) is also google translated as meteor, it is used thrice in the Quran to describe meteors by using several adjectives giving it this exact definition. And perhaps once for a torch of a burning brand in the story of Moses. But this is never used to connote a star. As such there was never any confusion among Muslim Astronomers concerning this important difference between a Star and a Meteor in the historic past.

> Quran 72:8-9 And we (jinns) had accessed sky but found it filled with stern guards (Angels) and **meteors.**
>
> And that we used to sit therein in stations for listening. But whoever listens now finds a **meteor lying in wait** for him.

The Arabic word شهابا (shihaban) translated Meteor is a noun praised by three different adjectives and active participles. To qualify the Cosmic definition of شهابا (shihaban) the Quran implies it as رصدا (rasada) stalking, stealth or lying in-wait [72:9]; then مبين (mubeen) meaning plain in chase [15:18]; then as السعير (as-sayeer) meaning a blaze [67:5] and then ثاقب (saqhib) meaning piercing bright [37:10] contouring erratic chase vectors unlike comets and large asteroids which rather traverse in more defined trajectories. In a verse 27:7 the Arabic phrase بشهاب قبس (bi-shihabinn qhabsinn) is used which translates as a torch for burning an object.[134] This word شهابا (shihabann) is used to depict the idea of a brand of fire which Moses set out to fetch when he saw a burning bush on the mount on a cold night. Thus, all meanings of the word meteor relate to an object that is reserved to alight such as space objects in traverse or otherwise relatively static like in Lagrangian points which when evicted are ablaze and surmount a clear

---

[134] Quran 27:7 (Mention) when Moses said to his family, "Indeed, I have perceived a fire. I will bring you from there information or will bring you a burning torch that you may warm yourselves."

chase in the regions of upper atmosphere torching the insurrectionaries. About the time in seventh century jinns were conversing that they weren't getting access to their traditional stations in space from where they would previously intercept classified information passing through first Cosmic sky. Per their conversations, the upper satellite regions of the earth were increasingly experiencing more severe meteor strikes. It is presently reported about 40 tonnes of meteor material burns up in the upper atmosphere each year. These meteor streaks in massive proportions aren't purposeless random events but objectively orchestrated activity for defined purposes. The world has witnessed it over the Russian territory as one of their resident jinns sped from a pursuing meteor in February 2013, of this chase many video footages are available on YouTube. Another daylight chase had been recorded on 22nd April 2012, a fireball as described was seen throughout the western United States accompanied by a loud booming sound which was heard across California's Sierra Nevada mountains. The Sutter's Mill meteorite a small asteroid, roughly weighing 88,000 lb was described as green and had a luminosity of magnitude -18 to -20 meaning that it appeared midway in brightness between the Sun and the Moon. It was reported by observers to be bright enough to dazzle the eye and subsequent analysis has found that it was made up of a rare type of carbonaceous chondrite seldom seen before which when heated results in oil & gas like Kerogens crediting the diversity of the Quranic description of word ونحاسٌ (wa-nuhas): *Quran 55:35 "Upon you be sent **space objects** of a fire brand and **of smoke**. And you won't be defended."*

Planetary objects, dwarf planets, asteroids, comets and meteors are not typical of our solar system. They are perhaps found across almost all-star systems in the Cosmic sky and are often undetected due to our astronomical limitations. Considering the Quran comets and asteroids are understood to take defined trajectories unlike meteors that eject out erratically upon crowds of jinnis speeding away. Orionids meteors- sporadic debris ejected from Halley's comet, take erratic paths exclusive for chase. With Space explorations being very arduous and expensive even for jinns to endeavour, they rely on solar-post based snooping for a channel to collect information close to home. This zone is therefore swelled in number of jinn operations. What ensues is increased counter-measures infused in the satellite skies of the sun and earth to reign in these illegal operations. Thus, many short and long period comets orbiting the sun shift trajectories slightly to align to be able to bulldoze reception networks of jinns.

And therefore, debris ejecting from comets like Halley's comet and others mount a barrage of meteors to incinerate jinn infrastructure. As the rebel comrades flee locked in a fierce blaze-chase upon entering earth's atmosphere deserting their fervently held posts for every such intersecting periodic orbital. Shooting stars is thus an expression, phrased as part of language to describe meteor streaks across the sky akin to phrases like "sun rises" and "moon set."

This differentiating narrative which classifies Meteors, Asteroids and Comets for their trajectories in space, strategy and make of materials cannot be abused as ancient folklore. Knowledge in the light of the holy Quran truly stands out and is the Light.

7.3 STAR ASTROSPHERES

Other than magnificently adorning sky with their presence, the Quran affirms that stars serve as guidance for navigation, barricading structures, as guards against intruding activities and as lapidation units to batter deadly radiation and arsenal ready to implode/explode.[135] This is Indicative of verses 67:5, 15:16-18 and 37:6-10 cited in this section.

> Quran 67:5 And We (Allah) have certainly beautified the sky of this world with <u>glowing lamps</u> (bright stars) and We have made them as deterrents for Satan-kind and have prepared for them the punishment of the Blaze.

> Quran 15:16-18 And We have placed in the sky <u>starry constructed mansions</u> and have beautified it for the observers. And We have protected it from every devil expelled. Except one who steals a hearing and is pursued by a clear meteor (blaze).

> Quran 37:6-7 Indeed, We (Allah) have adorned the sky of this world with an <u>adornment of bodies in whorls</u>—And as protection against every rebellious Satan.

> 8-9. (So) they may not listen to the higher assembly and are pelted from every side—Repelled; and for them is a constant punishment.

> 10. Except one who snatches (some words) by theft, but they are pursued by a meteor, piercing (in brightness, a blaze).

The singular word مصباح (misbah) means floodlamp or glowing lamp, translated in plural form in verse 67:5. In celestial etymology, star astrospheres flooding with brilliant brightness and deadly radiation over a wide spectrum in the energy band is a known info. This word **misbah** is found only twice in plural form in the Quran connoting specifically to stars and is tied to infer a fence to barricade

---

[135] Quran 16:16 And by (natural) mechanisms. And by the stars they are (also) guided. And, Quran 6:97 And it is He (Allah) who placed for you stars that you may be guided by them through the darknesses of the land and sea. We have detailed the signs for a people who know.

divinely sourced communication and further to give protection from the entities that are rebel some to God.[136] Critics misinterpret مصابيح (masaabiha) meaning glow lamps in verses 67:5 and 41:12 to act as missiles akin to meteors by wrongly reading the verse ending in 67:5: '*...and have prepared for them the punishment of the Blaze*', which in fact speaks of space embers (meteors, asteroids & comets) after explaining the role of stars in the first part of the verse. One of the awarded attributes for مصابيح (masaabiha) in verse 67:5 is رجوما (rujumann) meaning lapidation translated as deterrents which is further elucidated in verses 37:8-9 cementing the view of a constant punishment. The jinns & their operations are pelted constantly by the gamma-ray radiations and plasmas spewing in large chunks from more massive stars.[137] In Orion in a regional hot-spot where perhaps jinnis might be mounting operations to intrude find themselves overwhelmed by Betelgeuse; now a dying star and a ticking time bomb. Heartbeat stars and Pulsars further clarify the purposes of the given attribute رجوما (rujumann) as lapidation or pelting from every side.[138] Like the single vela pulsar with its astrosphere—pulsar wind nebula in the constellation vela, approximately 1000 ly away within first Cosmic sky is a supernova remnant and a millisecond pulsar, pulsating at every 89.33 milliseconds. This astronomical body spins approximately 11.195 times per second. The vela pulsar is noted as the brightest persistent object in the high-energy gamma-ray sky-scan. Pulsars and double pulsars are synchronised to beat very high-frequency pulses many times every second which delivers sustained battering across angular directions. Pulsars are a type of repellents which radiate two steady narrow beams of light in opposite directions. Pulsars can radiate light in multiple wavelengths, from radio waves all the way up to gamma-rays—the most energetic form of light known till date. The fascination is in Pulsar's light beam for lapidation that spins around like a lighthouse sweeping an area clear in a given plane and range. If the pulsar's light beam was to align with its axis of rotation owing to absence of God's Will and therefore wisdom; then it could not be credited to function like a repellent in a

---

[136] Quran 67:5 and 41:12 uses مصابيح (masaabiha) for stars in plural.

[137] CME's: The coronal mass ejections (CME's) is significant release of plasma & magnetic field from a star's corona. In such an event it travels to earth from sun in a span of two or three days and is shielded by the earth's Magnetic field.

[138] Heartbeat stars are binary stars (systems of two stars orbiting each other). It is said their brightness over time maps out like an electrocardiogram, a graph of the electrical activity of the heart. Hence the name heartbeat stars.

plane of operation as the beam could be a straight line ineffective for the purpose to function as guard. In explaining the word رجوما (rujumann) translated as deterrence via lapidation in verse 67:5 the incinerating radiation and lethal beams of light thrown off from bright star astrospheres and pulsars are being connoted as effective destabilisers that kill or terminate as the Quran is found to imply the word رجوما (rujumann) in criminal etymology used to kill by stones. Clearly many Prophets were threatened to be stoned to death for rejecting the cultural idol gods of their times one of them was Abraham as seen earlier:[139] *Quran 19:46 "(The father) replied: 'Do you hate my gods, O' Abraham? If you forbear not, I will indeed stone you to death...'."* There is a clear appraisal of the concepts invoked in verse 67:5 being elucidated in 37:6-10. In verse 67:5 the main restraint is coming from stars (glowing lamps) مصابيح (masaabiha) which are "**made as deterrents**" delivering the "**constant punishment**" by star-astrospheres, and then the later part of the verse talks of punishment of a blaze which is by a space ember. In explaining these verses 37:6-10 project the restraint to come from الكواكب (al-kawakib) which is a lumped model of all astronomical bodies working as a protective gear. By the stars of various kinds and of pulsars they are pelted from every side—Repelled as explained in *Quran 37:8-9 "(So) they may not listen to the higher assembly and are pelted from every side—Repelled; and for them is a constant punishment."* And by the non-starry and non-planetary space objects when a crime is cited there is -pursuance of a strike- by a comet, asteroid or a meteor: *Quran 37:10 "Except one who snatches (some words) by theft, but they are pursued by a meteor, piercing (in brightness, a blaze)."* Which does not always result in killing as in رجوما (rujumann) used in case of stars.

Hence all the three sets of verses from references 67:5, 15:16-18 and 37:6-10 first talk about deterrence by complete termination of intruders explaining lapidation whereas parts in connected context refer to a chase for a crime cited where the Meteors in pursuit don't always kill jinnis as they strive to evade the chase. More so in astronomical sense, the word رجوما (rujumann) is the catastrophic discharge

---

[139] The Quran hasn't stipulated criminal chastisement by lapidation of stones for adultery or for any other crime, period. Instead it talks of it as being used by people to persecute believers in the one God in the bygone days. Such as Quran 26:116 They said: "If you desist not, O' Noah! you shall be stoned (to death)." Also, see Quran 11:91.

of powerful radiations emanating from the star astrospheres. Star-implosions and pulsars that decimate or upset any approaching jinnis and their equipment are referred. The unswerving pulses of gamma radiation induce deadly inflictions and radio waves play jammer of the terrible order in jinn's abilities to listen—no pandering for them. It is like the spinning of high beams of light except that which they spew incinerating radiation instead akin to an army post built to spot and neutralize the enemy within range. But when the Quran talks of meteors in pursuit of jinnis as found in all possible references it uses the word تَبِع (tabaā) which means **following** or **to go after** as in a pursuit. It turns out that asteroids and meteors make an effective pursuant and comets ruin their infrastructure. Thus, we see the unfolding of purposes in constellations of stars, depositories of comets, asteroids and meteors being constructed in the **sky** of this world as deterrents for protection according to the Nobel Quran.

Stars on the outset makeup a massive Cosmic antenna which functions as a fence by jamming the incoming communication receipted by Angels at its outskirts. Indicative of a sharp context in verse 41:12 and further elucidated by a verse in *Quran 37:8 "(So) they may not listen to the higher assembly...",* other than for this type of quarantine stars make for fence & bombardment of jinn endeavours as discussed here. In verse 67:5 cited, analogous to the info in verses 15:16-18 and 37:6-10, the Quran projects the front-end of the stellar-design implying them as a beautification for the dwellers on earth but also to act as power zones for prohibition which builds security architecture to shield unauthorised disclosure of information inflow from Allah to Angels or between Angels. Considering the Quranic narrative, infantry of Angels positioned in the solar neighbourhood are on the lookout to quarantine satellite instrumentations of jinns in deep space by invoking astronomical adjustments. Angels function in limited ranges as none but God is the all-seeing.[140] This, therefore, necessitates large infantry of Angels to scout the interstellar space and in their approach near solar influence their numbers plummet due to reduced real-estate for their movement. One of the parameters invested in the design of the first Cosmic sky, the make of stars, solar system and its boundary region encompassed within the Oort cloud is to consider the scales of Angel's stately sizes to operate, who are distinctly upscale from our

---

[140] The Quran often invokes "all-seeing" as one among the many attributes of Allah.

general perception of sizes. Angels are huge but not all resemble the make of Gabriel, who is chiefly obeyed among their kind in the Milky Way domain.[141] However, for comfort of space to scout around solar system down to earth for their stately beings, if the organisation of solar system were not up to this vast scale they couldn't efficiently perform their work of monitoring rebel some activity of jinnis. It ensues, if we had the Angelic stately physical data plus knowledge of their work orders reverse engineering would aptly decode the scale and design from Oort cloud down to earth to what it is now.

We have seen that the Quran presents jinnis as smart and bright species who are much advanced in calculating celestial hotspots in space to station themselves. Particularly the L-points between two bodies or harness other unique methods of directional access of incoming signals. But there are obvious constraints, such as survival without essentials (plasma ions), balancing manoeuvres of their investments in the face of comets heading their way, managing speeds, sustaining infrastructures etc. This is in addition to the dangers posed by Angels to them. Their space endeavours become targetable by Angels whenever they attempt or successfully tap into a prohibited wire, at which point determinism is effected for the termination of their freewill as ordained in the law of God. The security procedures by the guardians screening for leaks activate the following troubleshoots upon detections:

1. Firstly, by vectoring orientation of multiple star systems such as binaries, triplets or other multi star systems summoned from an elusive dimension to gallop towards jinn establishments then to cause an inferno. Which is purely a time game between what jinns could manage to achieve before the galloping binaries intercept once reaching proximity.
2. Secondly, by effecting space objects for reconnaissance or chases for which Angels undergo momentary transfiguration from their elusive makeups to handle materials in our realm of space-time to surmount flames upon stealer jinns.

The latter is when Angels are spotted timely by jinns, surely not in their stately physical forms, but in forms suited for transiting into our dimension to exert physical force upon space objects like asteroids and meteors and dislodging them.

---

[141] Quran 81:21 Obeyed (by other Angels) and trustworthy.

This then culminates into fierce fiery chases of jinnis or their successful termination. On occasions chases could extend down to earth where infantries of Angels relay the chase procedures shifting the trajectories of space objects. Therefore, impending stern guards nearing Oort cloud to Kuiper belt and further down from asteroid belts eject meteors successively upon the fleeing Sabastian awaiting a grand reception from the rebel jinn communities living on earth. Recently surprising the scientific world, such a relay made news, you can read this on NGC: *For the first time in human history, an asteroid that originated from a different solar system has passed earth close enough for us to witness it. ... "It is going extremely fast and on such a trajectory that we can say with confidence that this object is on its way out of the solar system and not coming back."*[142]

An animation of this object's trajectory is available on NASA: *This animation shows the path of A/2017 U1, which is an asteroid -- or perhaps a comet -- as it passed through our inner solar system in September and October 2017. ... Designated A/2017 U1 – this object is less than a quarter-mile (400 meters) in diameter and is moving remarkably fast.*[143]

There have been many massive objects (comets) having close encounters with earth, apart from smaller objects (meteors or asteroids) crashing into earth's atmosphere. A/2017 U1 from a foreign location having a close encounter with earth is yet another reference. Time and again it is alluding of the chase that this was on and its purposes of protection alike many other comets still forecasted to tarry in closer trajectories to earth; in this case however the jinn is likely to have escaped this chase unless evidences of object revealed it was active (with flames) to torch the jinn.

Having read this far there is a tickler in the mind. Why does it require Allah, the Knowledgeable, Powerful Lord to orchestrate all of this? Why cannot Allah terminate jinn endeavours into space as they begin their missions from earth. Also, why don't Angels kill them right away? Or why must Angels in their real

---

[142] Title search: Mysterious foreign object tracked passing earth by Lulu Morris 30 October 2017 on nationalgeographic.com.au.
[143] Title search: Small Asteroid or Comet 'Visits' from Beyond the Solar System published on 27 October 2017 on nasa.gov.

forms be hidden from us and the jinns? If all the above claims were answered, then the mission of our test would end on earth. Allah is interested in evaluating us for our conduct and belief in Him.[144] Therefore, He has left open in His law the possibility for us to trespass limits on account of our freewill where the line of demarcation is His prohibitions.[145] Angels work on specific orders from Allah. When they are commanded, they plug astronomical blockades to control or let open the Cosmic grids to incite jinns into prying. In the law of Allah beings of freewill are tested for unfailing belief. There couldn't be a better test bed if we were to see Angels on stately forms, then we may freeze actions or be stunned on the token of it. Allah is far beyond any comparison even to Angels.[146] These are the matters of truth and Allah is determined to remain unseen from us in this life awhile we are being tested. Prophet Muhammad to whom Allah deemed fit to show Gabriel was so terrified after his first encounter with the Angel that despite Gabriel's greetings of salutations and Prophecy he shivered and sought for sanctuary asking his wife to shroud him. When Allah chooses a Prophet, He selects for able traits for the mission; which means they are worthy of it. I realised this in Makkah during my lesser pilgrimage, when I witnessed the diversified humanity from all corners of the world singing salutations upon Prophet Muhammad; singing in rhythms they extol his achievements. Allah's bestowal of honour is breathtaking! He was awarded prophethood for his demonstrated impeccable repute in his society for forty-years of his life being known as the most trustworthy, truthful and well-mannered individual. Until he opened the call to worship Allah alone from when he was accused!

> Quran 41:12 …And (also) We (Allah) adorned the Sky of this world with lamps as a **protective gear**. That is the determination of the Exalted in Might, the Knowledgeable.

The word حِفْظًا (hifzuann) here in verse 41:12 is translated as -protective gear-. It is drafted in accusative here and noted for no further contextual implications. Thus, it is interpreted as the back-end architecture of the interstellar sky making up the protective attenuation gear for the incoming divine radio communications.

---

[144] Quran 11:7 …that He (Allah) may test who among you is best in conduct.

[145] Quran 51:56 And I did not create the jinn and mankind except to worship Me.

[146] Quran 42:11 …There is nothing like unto Him (Allah), and He is the Hearing, the Seeing.

This word is also used in verses 37:7 and 15:17 as earlier discussed in this section. Stating specialized protective purpose i.e. protection from rebel jinnis making it something like the front-end part of architecture. The back-end architecture of attenuation is explained as follows: From the exterior edges of first Cosmic sky where star density is sparse which then gradually picks-up in its density towards our solar system, this interstellar star density comprises of hundreds of thousands of stars of common make (medium sized stars) in higher concentrations, then super novae at exteriors of this sky and in mid ranges, then multi-star systems then bright giant stars all akin to a network of radio-wave jammer circuit. They are like dense flocks of birds in all sequences of their life cycles: from massive giants to brightest main sequence stars, from heartbeat stars to ultra-high energy discharge pulsating-pulsars and many fainter star units most of whom are red dwarfs (dying). There are roughly 1400-star systems within 50 ly of volume of space containing 2000 stars surrounding our sun (**Figure 29**).[147] And in a volume of 2000 ly, approximately 80 million stars bubble per the same source. This network of millions and millions of stars in the solar neighbourhood (first Cosmic sky) spread across at least 3500 ly of volume designed for attenuating modulated band of divine communication is as elucidated in: *Quran 37:8 "(So) they may not listen to the higher assembly...."* It is evident from mere observation of its Cosmic photograph see **Figure 10**, **Figure 9**, **Figure 30** and **Figure 31**.

The sheer star density is built to perform as the jammer of the Cosmic wires broadcasted from Allah's ministries. Due to the Jamming effect the broadcasts can only be received by the angels stationed in the exteriors. That said, there are ranked Angels stationed in the Cosmic sky of our world from the peripheries of this sky going towards solar system and down to earth. These Angels in the interior receive the wires from the exterior Angels via line managers and report to Gabriel who then actions the affairs. Whatever of these signals which are not effectively quarantined by the star density, they simmer further edging towards solar neighbourhood where jinns tactfully position themselves to listen to these where the odds of jinns intercepting these wires increases. In the event of a successful or attempted intercept, Angels on duty begin their troubleshooting drills which is part of the front-end design facing the earth.

---

[147] Stars within 50 ly and 2000 ly on atlasoftheuniverse.com.

Gabriel is postulated to be stationed with his infantry very proximate to solar system around the Sirius binary system- the brightest known star system in near horizons. This subject matter is further discussed in section 7.6. Angels thus implement filed orders of Allah on earth directed by Gabriel who transmits pertinent teachings in the tongue of the chosen Prophets and Messengers of Allah. [148] Gabriel's sphere of information-connect to transmit with selected individuals is squarely specific. This connection is via the makeup that defines the individual by consciousness (Arabic: nafs) based identity. The Arabic انفس (anfus or nafs) meaning individual-self is the unfailing receptor akin to uniquely identified biological code. A make realised from information imbedded in the DNA, the quintessence of the surplus fluid emitted.[149,150] Upon identification of an individual Allah instructs Gabriel to make contact and teach whatever Allah approves of. Also, in the near regions of the earth the activity peaks as more often Angels must pass and communicate between themselves on many routines which they are told to bring into effect upon the earth exerting an influence on its dwellers. This is obvious from the range of anomalies that are evident in the first Cosmic sky and upon earth and in its skies, (**Figure 26**).

Thus, we see the timeless narration of the Quran expounding such perplexing astronomical concepts to us in this age. One needs to ask, what better or alternative words could the creator have chosen to explain the meaning to us which gestated a 1400-year period of scientific faceoff before an astronomically advanced age of 21st century arrived; where people are now capable of grasping that was said. This is indeed from the creator of the Cosmos!

---

[148] Quran 14:4 We send not a Messenger except (to teach) in the language of his people, that he may make (things) clear to them.
[149] Quran 86:6 (Man) is created from a drop emitted.
[150] Quran 32:8 And then made his progeny from a quintessence of a fluid despised.

7.4 Jinns—Intrusive Aliens

In the footsteps of orientalists criticising Islam, Fred M. Donner in his - Muhammad and the Believers- with praises from New York Times as "brilliantly original" has rethatched a long standing lie that Muhammad was influenced or borrowed ideas from near east (Judaeo-Christian) and continues to postulate that Satan (Iblis) in the Quran is the fallen Angel. Borrowing literature in 7th century to ready an impeccable book that shines for all times such as the Holy Quran to have been made from non-Arabic sources is a far simple impossibility. Firstly, because of those alleged source files being cited for many historical deformities. Secondly, it takes at least par excellence being a multilingual and a shortage of vision to postulate the Quran's future infallibility as a standing challenge for faith having closely criticised the wrongs found in those source files.[151] Beyond Judaeo-Christian modern Hindus make a similar claim of Muhammad copying "their" Vedic scriptures in this case one can imagine what calibre of super expertise Muhammad would have been mastering in the miserliness of Arabian desert having not had the means to write. If this were true, he alone is a better godhead than all of the gods, 'sons' of God & goddesses of the world together. However, Satan being a fallen Angel is a mistaken reading of the Biblical scripture by the church due to missing original scripts. After all, it is hard for Fred M. Donner to fight the delved baggage as he comes to think Satan in the Quran is also a fallen Angel.[152] It is vaingloriously disappointing of Sharmila Sen the executive editor at Harvard and of Mr. Fred himself for their claims made of contributions to intellectual treasures despite having had high proficiency in ancient languages such as the Quranic Arabic representative of Mr. Fred's book where falsehood is the premise of their publication. How much longer will the westerners rely on what seems to be "Nobel" institutions for learning Islam, while in their prestigious garb they only disseminate a bunch of lies and passions for misleading the world on a—truly Noble religion. Mr. Fred claiming that most people in Muhammad's believers' movement as illiterates and that they did not have a

---

[151] For example, incorrect writings of skies having pillars in Job 26:11 when alluded to by people were clarified in Quran- *Quran 13:2 It is Allah who erected the skies _without pillars_ that you [can] see...*
[152] Muhammad and the Believers, page 61; Harvard University Press, first edition,2012.

copy of the Quran (in a book format or on computer); in this case he is assertive that we can patiently comb it for problematic areas.[153] However, having avowed he has demonstrably failed to source it for the most famous verse that defines the origins of Satan and has thus combed a lie from thin air: *Quran 18:50 "And (mention) when We (Allah) said to the angels, "Prostrate to Adam," and they prostrated, except for Iblis (Satan).* **He is from the jinn** *and had departed from the command of his Lord. Then will you take him (Satan) and his descendants as allies other than Me (Allah) while they are enemies to you? Wretched is an exchange for the wrongdoers."* Rebels to Allah (Satan kinds) are from humans and jinns as we will see ahead. They are friends to each other in covering truth. So, will you take them as allies against the guidance of God? Therefore, here is a point of information that it is not the many Western or Islamic publications that will help you learn the clear truth, but it is only your critical self-analysis of the intact word of God— The Holy Quran.[154]

The question of Harvard press's proficiency in Quranic Arabic is realised here in this verse: *"Quran 18:50 "And (mention) when We (Allah) said to the angels, ..."* While Satan (Iblis) isn't an Angel so why then was the command even applied to him? This excluded bit of addressing Iblis here in this verse stands covered in the Quran by an inquiry with him concerning this matter. It is understood that Iblis was the communicator from God to Jinnkind akin to Messengers and Prophets among humans. When the command to acknowledge Adam was heard in this realm of Cosmic sky (our world) along with the many Angels stationed for orders as they are specifically called in the verse; Iblis a recipient of this communication stood obliged as he was later questioned for what had prevented him from acknowledging Adam's creation to which he answered: *Quran 7:12 "(Allah) said, 'What prevented you from prostrating when I commanded you?' (Iblis) said, 'I am better than him. You created me from fire and created him from clay'."*[155] In the Quranic narrative the one in authority and in majority is counted as the substantive for God's communications. Such as seen in many verses the masculine addressee based on masculine grammar do not free the feminine gender of its

---

[153] Muhammad and the Believers, page 77; Harvard University Press, first edition,2012.
[154] Holy Quran translation by Abdullah Yusuf Ali is recommended. Clearly, learning Quranic Arabic is the best way to stick to words of God.
[155] Also see Quran 38:75-76.

exhortations. Another classic among the prose is after humans were made inheritors on earth the jinns lost their status of addressee as seen in God's law. This is precisely why you'll see ahead, even jinns followed guidance of God from human messengers. And all God's exhortations to them were via specifics addressed to men as found commanded in the Quran. Though there are verses in the Quran addressing women folk and Jinnkind directly. Proficiency in the Quranic language is not about pointing at the adjectives and participles which any student is able at. But proficiency here must be demonstrated to understand its narrative substantive. As seen in the same verse the Quran declares that Iblis is of the jinn and not of Angels: *Quran 18:50 "... except for Iblis. **He is from the jinn ...."***

God informs us that Iblis and his descendants make an enemy to us whom we prejudiced stray to follow by their inducing of misguidance in us. It is belittling of an "educated" western society that in this day & age many come to opinion that among Muslims Muhammad is being considered as divine or at least on similar thought that most of them consider Jesus as God or the "son" of God. Among many other false attributions on Islamic heritage besides calling the seal of the Prophethood a war-lord by doomster westerners; we're told by simple googling (as believe many Hindus) that the black rock a relic of Abraham emblemed in one of Kaaba's corners in Makkah is the Shiv ling. Taken to be the sex organ of deity Shiva (also Siva) worshipped by Hindus.[156] The sheer size of the Universe alone by and large unambiguously declares that God is not a human with any of the animal organelles or features describing His existence. If not in 21st century O' mankind when will you learn the truth? It is appalling how we are being deceived! By rebel jinnis reducing God to manmade carvings. From the many prestigious institutions in the States to the sky scrapping towns and internet connected villages in India, in times of enlightenment with modern means lies are still easily drummed down many ignorant pals.

> Quran 55:15 And He created the jinn from (plasma) from **meadows on fire.**

---

[156] Linga (or ling) in Sanskrit is inferred to mean the sexual orientation of an object. To falsify this rendering, the popular school must redefine the Sanskrit word -ling- which is primarily used to identify the gender of a person by citing sex organelles.

Devil is a misnomer for evil ones among jinns, the rebels to God. Also, jinns cannot be mistaken for fiction-based devils depicted by artists as magical beings who can grant us any wish but with a hidden mischievous agenda, who shrink when trapped in bottles or old lamps, or as super powerful demons spewing fire in video games or vile beings synonymous with Satan. Jinnkind is a creature in plasma dimension on planet earth. They are made from مارج (marij) as per Quran 55:15, interpreted herein as plasma resulting from **burning of the meadows**. Traditional Islamic translations suppose it to be a mixture of fire or flame of fire or a fire without smoke.[157] Flames make the visible portion of fire constituting mainly gases at elevated temperatures via exothermic chemical processes of combustion. Substances of carbon dioxide, water vapor, oxygen and nitrogen constitute a flame; when hot enough in fire, they become ionised and turn into plasma. Plasma in fact is an appropriate rendering for the Arabic phrase—من مارج من نار (min marij min naar) translated here as **plasma from fire**.

It is known that Plasmas have many variations. Plasmas resulting from smokeless combustion of elementary substances are more difficult to detect. Much plasma known in universe doesn't emit in the visible spectrum of light. Even plasma in Plasmasphere and Ionosphere doesn't resonate with visible light and therefore light passes straight through it. It is likely that this is the stuff going into becoming the essence of jinn makeup. Materials that stay put such as sticky clay can deliver terrific dynamism when modelled into kinetic objects. Cheetah, a land animal has top speeds between 110-120 Kmh. Peregrine Falcon is among the top ten fastest cruising birds at 390 Kmh outracing Golden Eagle performing 322 Kmh. And in the water Black Marlin fish performs 129 Kmh. It is therefore understood that beings created from more dynamic forms of matter such as plasma (jinnis) and Angels from substance in an elusive dimension must necessarily sport superior abilities of displacements. Plasma unlike sticky clay does not stay put but is natured for high dynamism. The makeup of earth's essence derived from sticky clay are beings of flora and fauna, minute as microbes and large as mammoths. However, the makeup from the essence of plasma from fire are the Jinnkind. Intense fire and heat still makes a charge ready to incinerate

---

[157] Per Lane's Arabic lexicon page number 5800. Internet search for a pdf copy of Lane's lexicon.

them. It is indicated that they too have their kinds of beasts or vehicular modes of transport and board them like we back-board our race horses or vehicles.[158] However, the Quran has implied to their cruising speeds in relation to the meteor chases which they are involved with. Comets, asteroids and meteors have varying speeds through the space. The top speeds of space objects shooting in upper earth's atmosphere such as meteors is 28.6 km/s (64,000 mph) and those of comets on reconnaissance near solar heliosphere is approximately 70 km/s (1,56,000 mph). Though it is not clear whether jinns by themselves or by their beasts cruise such speeds, it is implied of old that they do evade such chase speeds. Thus, it is deduced that jinns travel on par or supersede the speeds of space objects. Also, by their technically crafted space ships they have launched major interstellar operations. It is likely that jinns have improvised great deal of techniques using star light. By momentum of photons called photonic propulsion for thrusting their plasma made ships into space. It means they do gas themselves into deeper space and back and forth perhaps in near volumes of celestial space. The stars within a volume of 10 ly from earth were touched even prior to Adam was tested. Wherein we understand Sirius's implosion to oust them surmounted a full-scale destruction of their interstellar platform. Which sent shock waves in their kingdom undoing stately arts sending them several hundred or thousands of years back into ancient life and their expertise severed due to deaths of many leading pioneers. Jinns since having encountered unprecedented challenges; therefore, it can be now postulated that their sole exertion here might be to inch on to critical grids toward the Orion nebula in their never-ending quest to pry. Such are the creatures of Jinnkind.

Indicative of the verses Jinns have anciently attained heights in technological progress and are much advanced than us. Innovations in their space endeavour are far superior due to their ages of lead over us. Thus, their ability to scale interstellar distances is not only plausible but it remains to be seen just how far they have gassed-up their explorations. Due to their make from plasma their dependence on oxygen ions for survival is derived. It is also clear from surah 72

---

[158] Quran 17:64 ...make assaults on them with your (Satan's) cavalry and your infantry...

that their scientific endeavours into space are anciently sought. However, their boundaries of reach are limited within near regions of first Cosmic sky.

Around the first millennium – light having travelled 6500 ly's of distance – the birth of Crab nebula was documented belonging to the Perseus arm another patch of sky in the Milky way. Since creation of the planet earth and post the rise of humanity it is likely that numerous other candidates such as Vela nebula within Orion spur in the Orion-Cygnus arm may have birthed suggesting that many star blockades effected for rebels on earth were unnoticed by us due to our long astronomical absence of sight and also our recent obvious limitations as light reaching earth from them takes hundreds to a few thousands of years to reach from within the boundaries of first Cosmic sky spread over a breadth of at least 3500 ly. It is construed that jinns live on surface earth and are most likely to swift through earthly skies by themselves or by their animals and dive deep oceans and move about every place from where they can source oxygen ions. Earth and its skies make a common cradle for us therefore our geo-celestial wanderings remain essentially the same. Like the satellite sky of earth-sun system or other planetary combinations and the interstellar regions. In the 7th century, the Quran records their scientific prowess whereby their capacity to launch into further regions of space augurs their technological advances to propel into orbits of the Lagrangian points as commonly known science among their kind. The zone of earth-sun based satellite space even for amateur jinns who aren't covertly committed to Satanic causes but in sheer exploration akin to human pursuits is a sure hot-spot. And many naive among jinns in their movements of discovering celestial sin perhaps may show us Meteor streaks rising sharply and occurrences of Comets and Asteroids in fierce action in near earth orbit. We can then imagine what is in stock with Satan's secretive old enterprise.

Allah in the Quran has narrated a conversation that concurred among jinns after they were directed to discover the Quran's revelation to Prophet Muhammad. Once the Prophet had just been through one of his most miserable days of rebuke and people's rejection while preaching faith. Due to heightened situations in Mecca he had headed to a town called Tuaif about hundred km from there in hopes of finding allegiance. But matters turned out for great disappointed as he faced utter rejection from the people of Tuaif towards his call. He was driven away bearing rebuke and stoned by children let lose merely for calling them to worship Allah alone. It was during his return journey that the spectacular event concerning the Jinn came to pass. Dejected at the turn of event in Tuaif, Prophet cried out to Allah in the loneliness of the desert. But Allah had ordained a lamp of upright Jinns pass by the vales and have them overhear Prophet recite the beautiful Quran. Details of their conversation amongst themselves after hearing the divine word is brought to awareness in the following pages:

> Quran 46:29 And (mention, O' Muhammad), when We directed to you a few of the jinn, listening to the Quran. And when they attended it, they said, "Listen quietly." And when it was concluded, they went back to their people as warners.
>
> 30. They said, "O our people, indeed we have heard a Book revealed after Moses confirming what was before it which guides to the truth and to a straight path.
>
> 31. O our people, respond to the Messenger of Allah and believe in him; Allah will forgive you your sins and protect you from a painful punishment.
>
> 32. But he who does not respond to the Caller of Allah will not cause failure (to Him) upon earth, and he will not have besides Him any protectors. "Those are in manifest error."

A continued conversation amidst them is related in chapter 72 in Quran.

> Quran 72:1 Say (O Muhammad), "It has been revealed to me that a group of the jinn listened and said, 'Indeed, we have heard an amazing Quran.
>
> 2. It guides to the right course, and we have believed in it. And we will never consider besides our Lord anyone (worthy of a Guide).
>
> 3. And that He, the exalted, is our ancient Lord; He is above being attributed to possess a wife or a son.
>
> 4. And that our foolish ones (peer jinnis) have been saying about Allah much enormity.
>
> 5. And we have come to opinion that mankind and the jinn should not speak upon Allah lies.
>
> 6. And there were men from mankind who sought refuge in men from the jinn, so they (only) increased them in burden.

7. And they (humans) had thought, as you (jinns) thought, that Allah would not resurrect any (for judgement).

8. And that we (jinns) reached the sky but found it filled with powerful guards and meteors.

9. And that we used to sit therein in stations for hearing: but whoever listens now finds a meteor lying in wait for him.

10. And we do not know (therefore) whether evil is intended for those on earth or whether their Lord intends for them a right course.

11. And among us are the righteous, and among us are (others) not so; we were (of) divided ways.

12. And we have become certain that we will never cause failure to Allah upon earth, nor can we escape Him by flight.

13. And when we heard the guidance, we believed in it. And whoever believes in his Lord will not fear deprivation or burden.

14. And among us are Muslims (in submission to Allah), and among us are the unjust. And whoever has become Muslim-those have sought out the right course.

15. But as for the unjust, they will be, for Hell, firewood.

From verse 72:6 cited here we come to know that men have been in contact with jinns since ancient times in various areas of interest. While the Quran is explicit in telling that men and jinn have attempted successful contacts by their endeavoured efforts, therefore the question of belief in this parallel dimension (concerning Jinns) is omitted. Schmoosing upon traders of faith since ages as they hold upper hand by working through kabbala influencing mankind to rebel it is conceivable that jinnis aren't still verifiable to us. It might take our science efforts to unveil their screens of hiding and turn the tables around. Perhaps here lay a determinism in our favour to begin control of them like Prophet Solomon could. The good folks among jinnis acting by determinisms mind their business by avoiding interfacing with us just like the animal kind that live sharing space yet do not interfere. But the rebel some are on missions to mislead men. They therefore intrusive interfere with us via secret arts in exchange for us declaring open rebellion against Allah. This is quite different from simply not believing in the Creator's existence. From parties involved in such allegiances what follows is lies, wilful obedience to Satans in all things wicked in addition to destruction of one's conscience—one of the separators between human and non-human. A logical alternative to this vileness is that we setup infrastructure to intercept Jinnkind yet hold our ground of leadership on them. If we sell our ground, they will deceive us as told in *Quran 72:6 "And there were men from mankind who sought refuge in men from the jinn, so they (only) increased them in (evil) burden."* However, our present inability to interface via techniques through technical instrumentation helps mystery surround any faith in such beings. Otherwise, we will be able to subject them in obedience under our command albeit not seeing naked eyed as our robust hand is laid on all other earthly creatures. The authority (technique) of sighting jinns was awarded to Prophet Solomon.[159] And he had jinns employed for duties- *Quran 21:82 "And of the devils were those who dived (oceans for pearls and artefacts) for him and did works besides that. And We (Allah) were of them a guardian."* The first temple which was destroyed by Nebuchadnezzar' II in 589 BC had jinn workmanship in it. Having said, it is very easy for every sane individual to grapple whispers coming straight from the jinns. This is perceivably felt when one is tested with decisions having to deal with embracing truth and justice. And more clearly felt when it comes to cynicism in the worship of Allah. And dominantly felt battling to quit sexual thoughts of promiscuous nature even as we work hard to be in control.

In the light of verses on jinn conversating a crowd of them after listening to the Quran reassured their faith in Almighty God recalling their latest adventures when they had taken to the skies, as they were staggered to see heightened activity by armies of Angels in their transiting and many newer fortifications filling solar interiors.[160] Asteroids and meteors occupying newer positions was a surprise phenomenon for jinnis as seen earlier in the Lagrangian points discussion in section 7.2. Therefore, the chieftain Satan and his troops weren't at liberty anymore even in familiar celestial spaces to conduct spontaneous operations

---

[159] Quran 27:17 And before Solomon were marshalled his hosts: jinns and men...

[160] Jinns can see Angels at the latter's transforming into dimensions to handle baryonic matter. Humans see them as strangers when they have attended to earthly missions, discussed ahead here and in section 7.4.

which is a contributing factor to their frustrations. They have increasingly seen their expeditions targeted and significantly narrowed down. They since have overly grown anxious to learn if Allah's declarations of wrath or mercy inches near to human inmates on planet earth. As the good buddy jinns seen in 72:1-15 resorted to renew belief in Allah they reckoned their mistaken ways to have become enlightened. Yet there are also others among them divergent and delusional indicative of verse 72:11. Unholy philosophy is clearly the whim of obstinate rebellion attempted to rationalise sciences favouring non-submissiveness to Almighty God. Humans and jinns in favour of freedom from any commitments to Allah resonate that Allah would not restore life again to nail us down for judgement which surely amounts to questioning -Allah's existence-. This perhaps brings timely peace of mind to such as those submerged in pride. However soon those reformist jinnis came to realise that they could never endeavour to cause failure to Allah here on earth nor by taking to flight in the Cosmos. God has only determined to gas our Cosmic explorations in ascertained limits to assess our freewill for belief in His existence. After overwhelming our consciousness by the of show universe and His sheer ability to host such enduring and magnificent creation artistry Allah promises to gather all of us back to him. Upon God's order an Angel posted nearby will sound the final horn and thence doom will ensue which will jolt the earth.[161] Awhile the catastrophe unfolds, it will engulf the first Cosmic sky and the Angels on orders will compel us and the jinns from our outposts from celestial space down to earth. It is very easy for Allah to snuggle all of us back to earth.

> Quran 7:27 O' children of Adam, let not Satan tempt you as he removed your parents from the garden, stripping them of their raiment of innocence to exhibit them sexually. Indeed, **he sees you**, he and his troops, from where (a dimension) you do not see them. Indeed, We (Allah) have made the devils partners to those who do not believe.

Jinns see us and can web contortions on to our minds from a dimension we cannot perceive them, as noted in the verse here. Their assaults are by implanting contortions via interferences over perceptions that we conceive at the level of

---

[161] Quran 50:41 And listen on the Day when the Caller (Angel) will call out from a **place nearby**.

neurons- a basic working unit of the brain. To this type of jinn's interference, the Quran refers it as الوسواس (al-waswasa), meaning whispers from the underworld. Jinns' only contention is to delude mankind from obedience to Allah in accordance with the promise undertaken by the chieftain of jinns due to Allah's honouring of Adam. While humans are tried consequently for our inherent nature to prove our mettle, despite jinns' bursting of ideas via bio plasma, we do have been awarded the ability to mastery over them which is by earning refuge in God via striving for truth and remembering God much.

> Quran 17:64 And incite whoever you (Satan) could among them (mankind) with your voice of assaults upon them with your cavalry and infantry. And partner them in their wealth and children and by promises. But Satan promises to them only delusions.

Adam's wronging before the command of Allah brought humanity to the fore of determinism for a long-standing mission of restraining from disbelief and disobedience. Therefore, at his slip Allah set out Adam from the innocence and conveniences of the garden of Eden to reproduce and strive on earth. Thus, humanity was forearmed for the great battle that life posed as a testing ground. With the token of forgiveness from Allah and a promise of timely guidance life for humans began on earth filled with tests.[162] And a new era for the receptors of freewill- men and Jinn began.[163] At this point many Astronomical modulations in the first Cosmic sky were effected to organise and secure the future of humanity and to straighten Jinnkind for their potential to interfering. Allah had thus brought about grand scale arrangements to safeguard the interests of humans from the subtle reach of jinns. Increasingly Allah has blockaded emergence of elusive sciences such as charlatans of Astrology, sorcery and magicians etc. Has this helped humanity? Rebels to Allah have always sought newer techniques to

---

[162] Concept of -Original sin- has no basis in the Quran. Islam teaches that sin is non-transferable and so is faith. Children do not inherit sin or faith from us. Men & women at maturity tick their respective clocks of sin.

[163] Beings with -freewill- are not the only ones with souls. This idea of only human beings possessing a soul is narcissistic. The "nafs" in humans and jinns is developed and modelled for freewill and accountability whereas in other creatures it is rudimentary and therefore not accountable for judgement.

dupe scores among mankind. Apart from cries of godlessness by Atheists to delude within the garb of faith popular still are deluders, curers, healers, number chanters, superstition etc. In the revelations of God, reliance upon Him alone is advised and we are asked to shun non-coherent arts such as Palmistry which is a form of divination. Nothing else but sure knowledge and verifiable procedures only must be inducted to conduct human affairs as it is the responsibility of the public state bidden by Allah to garner safe lives for men and women. Then for what good reason do foolish men and women blame God?[164] Truly a mind-virus of those sick who blame the Lord of the worlds for the folly of numerous-tumbling societies in pursuit of gain!

In a facet in Adam's disobedience to Allah concerning eating from the forewarned tree let us reason what would entail if it be that Iblis the chief Satan had obeyed Allah by acknowledging to honour Adam. Faced with the creation and honour of Adam if Satan had conducted in respect in his excellence, he would reach a true value for obedience to Allah in his determinism. Then in this case what would have possibly ensued for humanity? Determinism for both member recipients of freewill i.e. for jinn and men was organised or predestined as a decisive factor to test for truth and falsehood. All individuals of the family jinn and men are subject to respective determinisms to acknowledge or deny Allah. And to obey or discard His commands. The obvious consequence if Iblis were to be respectful of Adam's creation and honour of him ticking a true value before his determinism it would mean that Adam would still be directed to begin life on earth. And therefore, would undergo trials and temptations exclusive to him undeceived by Iblis's contortions on human mind. Which means the prohibition at the garden of Eden would not have been effective for Adam and Eve.

In the wake of Iblis's dislike of Adam and his disobedience to God it meant that Adam be educated about Iblis's challenge to lead humanity into disobeying Allah. Iblis's loyalty to Allah was on check at the design of Adam. Thus, Iblis came to stage of his logical-determinism which concluded with his refusal to obey. As he chose to exult before Adam denying him acknowledgement, it determined a false value in his test. Allah thus arranged to educate Adam of Iblis before

---

[164] Quran 5:90 O' you who have believed, indeed intoxicants, gambling, (sacrificing on) stone alters (to idol deities), and divining of arrows are but defilement from the work of Satan, so shun it that you may be successful.

Adam would embark on his journey in which humanity was to be assailed by Iblis and his kind. As a result, Adam's logical determinism was effected in which he learnt the knowledge of his enemy while in the garden of Eden.

Stars in the expanse of solar neighbourhood are for gaurding divine information from jinns owing to its functioning like a jammer of the modulated radio communications. The frequency of communications from high constituency broadcasted to Angels' reception is modulated for stealth in the cover of this spread network so that any prying instrumentation is overloaded by redundant radiation and radio-signals flowing from severely dense and other massive starry-bodies. Thus, jinns efforts for perception often experiences signal jamming of the terrible order. Even so, human intelligence can manage to differentiate starry noise for its very wide range in the EM spectrum but messages between Angels are vaguely continuous. Here is where the complexity is ever harder to make sense and is perhaps unfeasible if not implausible to decipher bits of divine information.

Jinns other than sighting Angels while they shift dimensions and as they scale down to tackle physical material, also sight them when they appear to aid armies of believers. Historically, it is told in the Quran that Allah, has ordered Angels to aid battles in favour of sincere believers making a terrific onslaught of the enemy.[165] So when the Angels transform to seek means to engage with humans in human forms, jinns are quick to take notice.[166] Where upon Satan turned on his heels from the battle of Badr from his efforts to bolstering the disbelievers. Badr is the first encounter of early Muslims against disbelievers of Makkah, in Arabia, fought at odds resulting in obliteration of the enemy lines when the enemy had mounted to quell Prophet Muhammad's communicated religion, not wanting to know, that it was from Allah. As the chiefs among the tribes couldn't see why Muhammad was favoured for prophecy over them, hence their pride of being chiefs had artfully positioned them to contend with Muhammad. A contest they really had to pay for!

---

[165] Quran 8:12 (Remember) when your Lord inspired Angels, "I am with you, so strengthen those who have believed. I will cast terror into the hearts of those who disbelieved, so strike (them) upon the necks and strike from them every fingertip."
[166] Quran 8:48 ...but when the two armies sighted each other, he (Satan) turned on his heels and said, "Indeed, I am disassociated from you. Indeed, I see what you do not see; Indeed, I fear Allah. And Allah is severe in penalty."

From this analysis we learn that the sixth century jinns efforts to tap information concerning human affairs was greatly affected as their trespassing now invited increased meteor streaks. Modern human efforts to build international space stations and launching of satellites is apparently very primitive to what jinns might be endeavouring in present times. Especially due to their sixth century learnings and the consequent restrictions enforced on them their advancements today must clearly be hundreds of years if not a millennium ahead of ours.

7.5   ANGELS—ELUSIVE ALIENS

> Quran 35:1 All praise is (due) to Allah- Originator of the skies
> and the earth. (who) Made the Angels messengers possessing
> wing (pairs), two or three or four. He increases in their creation
> as He wills. Indeed, Allah is over all things competent.

Cosmic spans are scaled by messenger Angels. They cruise with their
wings navigating across stars, nebulas, constellations and clusters of galaxies in
networked pathways. Their species are classed in great variation, and all of them
are equally elusive. As indicated *"Angels messengers possessing wing (pairs), two
or three or four"*, wings are added to magnify their stately form which alludes to
their great variation in sizes. In their elusive make-up and assortments, they
accelerate to speeds beyond current human understanding. Travelling at speed of
light (c), Photons reach earth from Sirius in about eight and a half years' time.
Whereas Angels in their elusiveness are on instant speeds from the star Sirius as
they manifest on earth when needed by transfiguring into human forms instantly:
*Quran 19:17 "She (Mary) took seclusion from them; then We sent her our angel
(Gabriel), and he appeared before her in the form of a man."* It seems there is a
subtle science to grab here. When Angels reach the earthly firmaments (skies)
they're imbued with abilities as required to sly into anthropomorphic forms from
their Angelic makeups. As told earlier transformation also occurs in outer space
when they manoeuvre space objects to overwhelm Jinn kind giving them a
window to be spotted. Regardless it is an ability given to them to transfigure at
will. Angels move in elusive speeds when they travel between star posts or while
travelling via Cosmic pathways in their stately forms but when they interface
with material tokens such as us or jinnis requiring a transformation, they scale
down to relative abilities accordingly. Gabriel for his dealings with the nature of
prophecy in human history has made several quick descends from within first
Cosmic sky from his post at Sirius to the earth. It can be said that the travel time
was almost instantaneous. So also have the armies of Angels descended from this
barrack-post to aid believers.[167]

---

[167] Understanding of speed and hangout derived from references in Quran 3:124-128,
8:9-14 and 86:4.

Angels travel through universe via the Cosmic highways descriptive of the Quranic information which is not analogous to the imaging of Cosmic Wormholes.

> Quran 70:4 The Angels and the Spirit (arch Angel Gabriel) ascend to Him (Allah's throne) in a Day which has a measure equal to fifty thousand years.

As a point of information in the Quran, Gabriel is invoked specially besides the verses invoking Angels collectively as seen here in verse 70:4. Gabriel is the arch Angel. He is bestowed with the title Ar Ruh—Al Khudus translated as The Spirit—The Holy. Chief of the infantry in this domain of Cosmos and Honoured with God for endurance: *Quran 53:5-6 "(Revelation) Taught—by him (Gabriel) intense in strength. One of soundness (and wisdom) ...."* Pondering over the Quran explicates many exciting revelations which are recorded in the coming pages. Tellingly Angels' speed is alluded to in frames of time. This information is entwined in more than one place as already seen in verse 19:17 and also described in good detail in battle circumstances. We have seen in section 5.3 a couple of Angels per person are given a rank in the Sirius arena: *Quran 86:4 "There is no soul but that it has over it- guard."* They are made godfathers over us recording details of circumstances ahead of us then noting down our actions for what we commit in response to given situations and nonetheless guarding us unless where the Will of Allah prevails: *Quran 13:11 "For each one are recording (angels) before and behind (i.e. by before it means upcoming circumstances and behind it means committed actions) guarding him by the decree of Allah. Indeed, Allah will not change the condition of a people until they change what is in themselves. And when Allah intends for a people ill, there is no repelling it. And there is not for them besides Him any patron."* When determinisms[168] from Allah approach such as coming of a Messenger (for example Noah, Abraham, Moses, Jesus and Muhammad) advocating God's specifics, eventualities lurk very close: *Quran 8:13 "...if any contend against Allah and His Messenger, Allah is strict in punishment"* then in this case if such a Messenger is forced to defend, Allah backs His Messengers with support by sending Angels from their barrack posts in Sirius with instructions to kill concluding their determinisms: *Quran 8:12 "(Remember) when your Lord*

---

[168] Determinism: Also a decree of God to interfere into human affairs at the level of wider population. It mostly concludes in death & destruction and can also result in suffering and losses. See section 9.2.

inspired the angels, 'I am with you, so strengthen those who have believed. I will cast terror into the hearts of those who disbelieved, so strike (them) upon the necks and strike from them every fingertip'. " In battle situations when prayers were invoked by Prophets, God honoured them by sending Angles to timely aid the humble believers. This is a definite indicator to put an estimate on their velocities. It is perceptibly a journey covered in a matter of seconds perhaps instantly from the locus of Sirius's deck to the battle grounds on earth and which per evidences from the Quran occurred just in time during the battle causing a massive onslaught of the enemies. Like in a reminder rehearsed in the Quran after losses in the second battle of Uhud where a band of archers did not keep in obedience to the Prophet: *Quran 3:124-125 "(Remember) when you said to the believers, "Is it not sufficient for you that your Lord should reinforce you with three thousand angels sent down? Yes, if you remain patient and conscious of Allah and (in future) the enemy rush upon you by a quick manoeuvre, your Lord will reinforce you with five thousand angels having marks (of distinction)."*

Before, in the plains of Badr (first battle), as the enemy came in sight of Prophet Muhammad, he was overwhelmed seeing their numbers. He fell on his forehead pleading to God for help as the humble believers had sacrificed everything for faith and had already endured thirteen years of persecution in Makkah at the hands of these polytheists: *Quran 8:9 "(Remember) when you asked help of your Lord, and He answered you, 'Indeed, I will reinforce you with a thousand from the angels, following one another'."* Then as the armies faced each other for a bout: *Quran 8:48 "…but when the two armies sighted each other, he (Satan) turned on his heels and said, 'Indeed, I am disassociated from you (disbelievers). Indeed, I see what you do not see; indeed, I fear Allah. And Allah is severe in penalty'."* As this milestone battle raged, believers firmly holding ground proved their unwavering faith to Allah; thence God fulfilled His promise and armies of Angels with instructions to kill were realised amidst battle forces. This clearly alludes to Angels quick velocities for instantly covering great distances. Though there is a substantial time interval between God's committing of heavenly forces in response to Prophet Muhammad's prayer and Angels arriving at the Badr plains, it does not impede Angels having swift speeds. When Satan saw this metaphysical phenomenon, he fled from the ranks of disbelievers and Allah then filled fear in their souls, thus paving way for their defeat. Prior to the arrival of Angel units, the Prophet's adversaries had seen his army as a small and contemptible force which drove them in enthusiasm to crumble it. In almost all the battles which the Prophet fought his forces were at odds. Muslims were compelled to fight with far lesser numbers and were marginally equipped, except during the liberation of Makkah. During the Makkan siege compassionate Prophet Muhammad had granted general amnesty to citizens of Makkah. This was his moment of return from an eight-year exile and an unparalleled moment of kindness in history of mankind.

In the very next year after Badr the battle of Uhud was fought wherein the disbelievers seeking vengeance had formed a massive confederation bringing with them several tribes from the recesses of Arabia and marched up their horses to the doorsteps of the city of Madinah where Muslim families had their residences. Anxious of an imminent attack Muslims had camped at a nearby hill called Uhud. In this case Prophet Muhammad summoned the believers with orders for battle reminding them of how God had helped the faithfuls in their first bout in Badr descriptive of the verse: *Quran 3:123 "And already had Allah given you victory at (the battle of) Badr while you were few in number. Then fear Allah; perhaps you will be grateful."* The battle was in favour of the Muslims initially, but with the imminent victory at sight, a portion of the Muslim army slacked and a rank of archers posted by the Prophet to secure Muslims position left their positions seeking after the booty. It was then that God withdrew His supports reproaching them for their indiscipline and travesty to the Prophet's command: *Quran 3:152 "And Allah had certainly fulfilled His promise to you when you were killing the enemy by His permission until when you lost courage and fell to disputing about the order [given by the Prophet] and disobeyed after He had shown you that which you love. Among you are some who desire this world, and among you are some who desire the Hereafter. Then he turned you back from them [defeated] that He might test you. And He has already forgiven you, and Allah is the possessor of bounty for the believers."* Then the divine reprimanded was followed by Allah consoling them: *Quran 3:125 "Yes, if you remain patient and conscious of Allah and (in future) if the enemy **rush upon you by a quick manoeuvre**, your Lord will reinforce you with five thousand angels having marks (of distinction)."* This is at best informing of the time interval needed for Angels to descend for aid. In their haste to kill even if the disbelievers had rushed upon Muslims, this interval was sufficient for Angels to cover their journey from Sirius to earth for aid. Even if we idealise a new scale where a light-year's worth distance is travelled in one second, then it would take 8.6 seconds for Angels to reach earth from Sirius as Sirius is 8.6 light years away. Now going back to Quran 70:4 we are given the time taken by angels to reach Allah's abode as 50,000 years of our reckoning: *Quran 70:4 "The Angels and the Spirit (arch Angel Gabriel) ascend to Him (Allah's throne) in a Day which has a measure equal to fifty thousand years."* Calculating the number of seconds in 50,000 years, this approximation scales the size of the universe in a neat vector pointing from the Milky Way to the Throne of Allah in His abode to be more than 1.5 trillion light years (ly) across. Based on this finding, the present estimates of the observable universe said to be 46 billion ly in breadth is truly an atom (warning here of an exaggeration!) to the universe spread across clearly greater than 1.5 trillion ly as this measure is only in a given straight vector![169]

Angels' make is from an elusive energy not revealed in the holy Quran. This is because Allah has deemed it to be of matters related to the unseen. As we know that only a small band in the electromagnetic (EM) spectrum is in the visible range. Though more of EM is detectable and as a result is digitally mappable, therefore, making it vision-able. But besides this know, the detectable band of EM; there are many unknowns which are understood to exist. It is not known of which elusive substance Angels are made from although it is alleged on behalf of Prophet Muhammad, that they are made from نُور (nur) translated as glow of light, implying photons. Being a revelation, the Quran is the primary source of truth. If this detail isn't mentioned in it, then in preference to concreting views based on an alleged report or spurious readings of Prophet's speech, we need to look again in the Quran for why it has omitted such a detail. The scripture calls for belief in **Angels' existence** an exclusive call missing say for jinns.[170] This is

---

[169] Pardon me and the scientific community for exaggerations. In popular science, scientists have predicted this comparison; Cosmic Journeys- The Age of Hubble by Thomas Lucas Productions, Inc.; published 16 April 2015 on YouTube; time: 18:27-19:00 minutes.

[170] Quran 2:177 It is not righteousness that ye turn your faces towards east or west; but it is righteousness- to believe in Allah and the Last Day, and the Angels, and the Book, and the Messengers; to spend of your substance, out of love for Him, for your kin, for orphans, for the needy, for the wayfarer, for those who ask, and for the ransom of slaves; to be steadfast in prayer, and practice regular charity; to fulfil the contracts which ye have made; and to be firm and patient, in pain (or suffering) and

because Angels are elusive and cannot be contacted, except by Allah's exclusive grant to those of men whom he chooses for revelation or as they become permeable at the final moments before one's death.[171] Or for trial as seen in Solomon's kingdom (talked early in chapter 7): *Quran 2:102 "And they followed (instead) what the Evil ones spelled in the reign of Solomon. It was not Solomon who disbelieved, but the rebels disbelieved, teaching people magic and that which was revealed to the two <u>angels</u> at Babylon, Harut and Marut. But the two angels did not teach anyone unless saying, "We are a trial, so do not disbelieve (by practicing magic)." And (yet) they learn from them that by which they cause separation between a man and his wife...."* Belief in their existence is thus explicitly stipulated and is aptly made an article of faith.

Angels however do bless tidings by interfacing with humans perhaps via the bio-plasma channel (which is surely used by jinnis but to murmur deception) or are entirely elusive by imbuing sensationless tinges upon the righteous among believing humans saying: *Quran 41:30-32 "Indeed, those who have said, 'Our Lord is Allah' and then remained on a right course- the angels will descend upon them, (saying), 'Do not fear and do not grieve but receive good tidings of Paradise, which you were promised. We (angels) are your allies in worldly life and (are so) in the Hereafter. And you will have therein whatever your <u>souls</u> desire[172], and you will have therein whatever you request (or wish). As accommodation from a (Lord who is) Forgiving and Merciful'."* However, the Arabic word يوحي (yu-ha) inspiration or وحي (wahi) upon the conscious mind of a person is not employed in verses 41:30-

---

adversity, and throughout all periods of panic. Such are the people of truth, the God-fearing. And,

Quran 4:136 O ye who believe! Believe in Allah and His Messenger, and the scripture which He hath sent to His Messenger and the scripture which He sent to those before (him). Any who denies Allah, His angels, His Books, His Messengers, and the Day of Judgment, he has gone far, far astray.

[171] Death in sleep (even for disbelievers) is the easiest an understanding derived from Quran 39:42 and Quran 6:60-61. As the 'nafs=the conscious self' is already retrieved for the purpose of sleep therefore angels do not encounter people in this case to chastise or greet but God Himself further severs the soul to result death.

[172] The translated word soul in this phrase like in many other verses is not equivalent of the Arabic word 'ruh' i.e. 'spirit' but it is equivalent of 'nafs' which means the self, the mind or simply said consciousness which is written as soul here for language purposes.

32 here clarifying that the divine direction is not the premise.[173] Angels do voluntarily descend & overlap upon the consciously persevering faithfuls seeking prayers of God for them. This could be explained perhaps as the bio-plasma interface from wherein Angel beings would ooze to human beings or an equally elusive technique seeking a channel to convey us tidings that we realise as perhaps good feelings or hope while we earnestly pray. Therefore, the verse rightly tells us of Angels' descent upon earth and their being amidst us to herald comforting inspirations. Going by this premise God too inspires human minds of numerous enquiring men regardless of prophecy towards subtle truths in all of our respective endeavours such as the idea occurring to Newton for his postulates in fundamental laws of physics.

Thus, for real time interface where conveying a revelation is not the premise the Angels draw near to the believers reassuring them to endure and persevere faith. But for وحي (wahi) as in for a revelation of a scripture a learned Angel possessing wisdom is chosen for the task. Therefore, Gabriel- the holy spirit delivers it to an identified human recipient capturing the present, past and future scenes in the revelation (see section 5.2).

This revelation having been given to Gabriel in the form of codecs of laws despatched from God via radio-communications is understood and then transmitted to a series of human recipients in their native tongue by Gabriel:

> Quran 2:97 ...it is (Gabriel) he who has brought the Quran down upon your heart (O Muhammad), by permission of Allah...

...the set of neurons in the brain dealing with the heart is made the receptor, and

> Quran 53:5-6 Taught (Muhammad)- by him (Gabriel) intense in strength. One of soundness (and wisdom) ...

---

[173] But the word وحي (wahi) is used in several non-scriptural contexts: 16:68 for guidance to bees collecting nectar; 6:112 rebel jinn and men (subconsciously) promising each other delusions; 6:121 Satan murmuring to likeminded fellows; 19:11 signals via body language; 21:73 general guidance to all men towards goodness and progress; 23:27 inspiring ship design & construction and 41:12 inspiration to the skies & earth in the context of non-living entities; etc.

Gabriel has also presented himself in human form for many interactions with us, such as in the story of Mary the chaste, giving her tidings of Jesus Christ in her virgin opulence:

> Quran 3:47 She said, "My Lord, how will I have a child when no man has touched me?" (The angel) said, "Such is Allah; He creates what He wills. When He decrees a matter, He only says to it, 'Be,' and it becomes.

So also did Gabriel tread amidst the group of Men when Prophet Muhammad with his fellow believers was reclining in remembrances of God. He appeared immaculately presented which evaded any signs of travel or fatigue coming from this stranger and he came to be seated touching knee to knee with Muhammad and asked him a few fundamental questions concerning faith. Then when Muhammad presented his answers the stranger confirmed them and left. Upon enquiry by his companions of this immaculate stranger, the Prophet informed them that it was Gabriel who had come to teach them faith.

In the first interaction with Muhammad on mount Nur in the outskirts of Makkah, Gabriel exchanged words after greeting him as the Messenger of God. All this were real-time interactions not hindered by time dilation. This means, in real-time communications including revelations of scripture there is a subtle science of propelling waves (signals). We understand photons to have a velocity c but if these photons were propelled, they set new standards for speed. Like EM radiation travels at variance in different mediums such as vacuum, water etc. Angels from an elusive dimension are able to manipulate electromagnetic (EM) radiation. Thus, their communications are starkly at variance from the general EM spectrum we receive nearly at the speed of light from the Cosmic background and starry influxes. This also explains Gabriel's ability to communicate in real time as with Muhammad's first interaction and also to give scripture from a distance from Sirius without dilation because radio communications needing spans of time delay to reach earth from 8.6 light years would not be able to achieve this. The speed of Angelic communications is given by the equation in verse 32:5 *"He arranges matters from the (earthly) sky to the earth; then its (reporting) will ascend to Him in a Day, the extent of which is a thousand years of those which you count."* Thus, back calculating and factoring the distance of 1.5 trillion ly from Milky Way to the Throne of Allah with the number of seconds in 1000 years we get half a billion million km per second of speed for punched wires by Angels. This number factored with the speed of light c gives us the speed of communication to be a billion times faster than the speed of light. Amazing as it is it is not surprising if you think a speed of this magnitude is required to be able to deal with the degree of expanse presented in this vast Universe. Reporting of events or audits of accomplishments upon planet earth are packaged by Angels to be wired to God. This wiring technique ascends the information of affairs to God in a day equal to one thousand years of our count. God is all aware and knowledgeable as He clearly informs us. The reason why Angels post them is for the science of evidence keeping and for facilitating knowledge for the ranks of Angels with whom God convenes many matters of worldly affairs as Angels are given to pray for better living of humans. And as these guys become the information bearers and senders to the world down under knowledge and awareness of worldly affairs is for them an essential belt. If God were to unleash determinisms upon earth and not produce evidence, we would remain doubtful of our punishments on earth and in Hell, despite God's all knowledgeableness and His attributes of justice. If evidence is shown we are given to see every account of our keener participation in rebellion to Him, then surely it would restore us full regrets for our actions committed (see section 9.1). Determinisms on earth in a way differs to personal judgements on doomsday for which individual evidences are kept with Angels homed at Sirius who will hand us over our records on that day.[174]

Nevertheless, by this relation derived by equations of a day for Angels to travel to God as equal to 50,000 years of our count and the information travel to God as equal to a day of 1000 earthly years; the information travel time is straight 50 times faster than Angels' travel time. This squares Gabriel to communicate seated from Sirius to chosen men on earth in their prophecy without any impedance to communicate in real time. Besides this if Gabriel were to communicate from Orion nebula descriptive of بالأفق الأعلى (bil-ufqhil-āla) in verse 53:7 instead of from Sirius it would mean that factoring the distance of Orion (1300 ly) from us by 50 there would be at least 20 seconds delay for Gabriel's communication to reach earthly recipients in matters tied in real-time inquiry. Our voices are heard up to a wide

---

[174] Determinism: Also a decree of God to interfere into human affairs at the level of wider population. It mostly concludes in death & destruction and can also result in suffering and losses.

range in the near celestial sky by the Angels as they listen to us by the quantum bells for our voices perhaps transit via the Quantum modulations. As a result, Angels have positioned in the near deck in Sirius for divine communication and recording activity instead of afar from atop the stage in Orion. Thus, what ensues is the star Sirius is making an ideal Angel-post for communications and security issues.

Gabriel owing to elusive abilities could manipulate photons and become permeable to select locus positions for being visible to the target alone such as say only to Muhammad on that historic day at Mount Nur. The second time Prophet Muhammad saw him was when the Prophet himself was in the horizon of planet paradise outside the Orion spur, the cocoon of solar neighbourhood. Yes, this is the renowned night journey of the Prophet where the Quran reveals that he was transported to the paradise planet: *Quran 53:13-15 "And indeed he saw him in yet **another descent**. At the prohibited Lote-Tree. Where there is—the enduring gardens."* This topic is covered exhaustively in section 7.6. Now, as if it is a modern miracle unfolding before our very eyes when the Quran talks of paradise planet and Prophet having been taken there on a night journey.[175] The Quran specifically talks of **yet another descent** of Gabriel in that locus of Cosmos. The miracle of the Quran is thus proved astronomically. If the Quran were not to talk of **another descent** of the Angel in the realm of a different patch in the higher Cosmic sky this would categorically question the Quran's deliverances on celestial sciences. The dive of the Angel is incumbent as the locus of paradise planet is vastly further from the cocoon of the first cosmic sky, we live in. Having given such a detail, it reassures the Quran's divine origins for its deep elucidation of Cosmic facts 14 centuries ago. Alas! for the naturalists Cosmos from far is thundering.

Further, vision is facilitated only when light is incident from a dimension therefore the staggering abilities of Angels to lens their stately image from that great distance is an amazing feat. Human organ for vision perhaps is more capable than what has thus far been realised. As and when Allah allows, we become vision-able to see the unseen. However, it could be argued that the data of Angel Gabriel's stately appearance was transmitted to Prophet Muhammad's

---

[175] Via space-time travel.

conscious mind akin to him receiving revelations making him feel that he has grasped the sight of the Angel. Therefore, it can be dissented that Muhammad actually didn't grasp a sight of him via his eye organs, but the image data was transmitted to his mind like a revelation. However, a few verses from surah an-najam, -the star- clarify the Quran's narrative as follows: *Quran 53:11-13 "The (Prophet's) senses did not fault that which he saw. Will you then dispute with him over what he saw? And-Indeed he saw him in yet another descent"*, verses herein simply refute the ill-conjectured position and then confirm Muhammad's actual sighting of Angel Gabriel via his eye organs.

Angels live for aye; Allah has let nothing in the Cosmos to hurt or terminate Angels. They do not become jarred by fires or say warped by blackholes.[176] Which means star astrospheres and other space objects when ablaze will make definite deterrents to repel jinns but not for Angels. Jinns along with humans share the burden of responsible freewill and therefore judgements and repercussions in hell over which are appointed nineteen Angels as guardians: *Quran 74:28-30 "It lets nothing remain (unburned) and won't spare. Scorching them thoroughly. Over it are nineteen (angels)."*

Angels are responsible for directing anomalous weather and also distribute the sustenance of Allah upon the earth as per the providence plan. Natural patterns of seasonal winds, distribution of fecundating winds, winds bearing beneficial rain clouds or otherwise and many anomalies such as cyclones are spearheaded by Angels for providence and importantly for humans to learn compassion and righteousness accordingly. The gusts are thus driven to cities as justifications or as warnings for people to suffer for committing to oppression or wrong actions in the law of God: Quran 77:1-7 *"By those (winds) sent forth in gusts. And the winds that blow in violent rage. And (by) the winds that spread (distributing clouds). And those (angels) who distinguish (winds). And those (angels) who expedite instructions. As justification or as warning. Indeed, what you are promised (judgement day) is to occur."* As seen here in these verses various Angels are appointed for handling different kinds of winds. Sustenance and water resources are huge providential assets which are thoroughly regulated by God to provide

---

[176] Quran 7:20 ...he (Satan) whispered: "Your Lord only restricted you this tree, lest you should become Angels or such beings as **live for ever**."

for the many inhabitants of the earth. Angels distinguished in this business expedite orders from God. Gabriel himself with his armies makes it to earth for specific affairs once every year: *Quran 97:4 "Therein (in a night in Ramadhan, an Islamic holy month of fasting) come down the Angels and the Spirit (Gabriel) by Allah's permission, on every Errand"*, to detail plans for another year.

> Quran 35:44 Have they not travelled through the land and observed how was the end of those before them? And they were greater than them (Quraysh) in power.[177] But Allah is not to be caused failure by anything in the skies or on the earth. Indeed, He is ever Knowing and Competent.

> Quran 29:22 And you will not cause failure (to Allah) upon the earth or in the skies. And you haven't got other than Allah any protector or any helper.

If you could search for -Lost city- on Wikipedia: There are numerous lost cities ranging from Africa's rediscovered legends, cities lost in Asia, central Asia, far east Asia, south Asia, south east Asia, western Asia/middle east, Russia, Europe's twenty-three countries including UK, North America including Canada, Mexico and central America, South America and others. The ruins of former civilisations lay desolate. Besides the fate of Noah's people, the Quran in its revelation to Arabs brought their attention to naming a few in their immediate vicinity: The Aad of Palmyra—popularly known as the bride of the desert (Hud was last among the Messengers sent to warn), The Thamud of Petra—popularly known as rose city (Saleh was last among the Messengers sent to warn),[178] ruins of Shuaib in the Madyan (Jordan region), the ruins of Prophet Lot (Dead sea region), Egyptian Ramses II, affluent of Pharaohs (Egypt region), attention is also drawn to Pompeii in Quran 36:13-29 and Indus valley Harappa and Mohenjo daro indicative of 34:15-17 and other cities between them indicative of 34:18-21. Then, more newer paradigms were admonishingly presented to instruct the new

---

[177] Quraysh: Muhammad belonged to the tribe of Quraysh from whom God had then selected a Prophet in Mecca (Makkah).

[178] Quran 41:13-14 But if they turn away (from the Quran), then say, "I have warned you of a thunderbolt like that which struck Aad and Thamud. When the Messengers had come to them in succession, [saying], 'Worship none except Allah'. They said, 'If our Lord had willed, He would have sent down the angels, so indeed we, in that with which you have been sent, are disbelievers'."

community of Muslim believers learning historical wisdom around Prophet Muhammad. The destruction of Solomon's temple in Jerusalem by Nebuchadnezzar II in 589 BC in the wake of numerous sins of Hebrews and due to them violating the solemn covenant ratified with the Lord of Moses; followed by its second destruction by Romans in 70 A.D. in the wake of Israelites rejecting of Prophet Jesus Christ the son of Mary, a chaste Hebrew. Allah's law in human sphere works for belief and righteous conduct. If civilizations are found in violation of belief by scrapping consciousness of Him and run amok in delusions of free reign, destroying order, he therefore replaces them. In yet another law, despite belief if a believing people neglect morality and fail to rise above oppression, impunity is not upheld by Almighty Allah. Allah punishes by His unique means.[179] Allah's retribution will also surround a believing community if believers lax and quit working towards upholding humane values towards one another and rights towards the creator.[180]

> Quran 10:47 And for every nation is a messenger. So, when their messenger comes, it will be judged between them in justice, and they will not be wronged.

In the law of Allah, injustice and oppression has a measured time line. When the appointed time lurks, Allah sends messengers who declare His truth and enforce amends. If the society responds willingly, Allah blesses them even more, but if they chose to exult and oppose, Allah replaces them with newer generations. Thus, He directs Angels to specific tasks who precisely perform what they are bidden to, being stationed near our domains unnoticed as they go about their businesses. They channel storms of space weather or bring in storms from atmospheric weather or bring down meteors and stones from exterior boundaries

---

[179] Quran 6:65-67 Say: "He (Allah) has power to send calamities on you, from above and below, or to cover you with confusion in party strife, giving you a **taste of mutual vengeance** - each from the other." See how We (Allah) explain the signs by various (symbols) that they may understand.
66. But your people reject this (Quran), though it is the truth. Say (O Muhammad): "Not mine is the responsibility for arranging your affairs."
67. For every message is a limit of time and soon you will know it.
[180] The world has witnessed this law of Allah unfold within communities of Jews of the old and then Christian communities of lately times and presently in the Muslim world.

or even turn pockets of the earth upside down for they are obedient, mighty and unfailing in their tasks.

> Quran 7:34 And for every nation is a (specified) term. So, when their time has come, they will not remain behind an hour, nor will they precede (it).

Globalisation is a clear indicator for the culmination of a grand show on earth. Because it has made collaboration possible across humanities and has realized our endeavours towards common objectives. The knowledge concerning wonders of Allah's artistry are shared entirely. Now humanity is culpable in a common thread for the flow of ideas concerning our origins, purpose and models of events following death. In the event of overwhelming evidence of creation, if sections of humanity aren't persuaded from disproving the plausibility of Allah's existence and further obstinately deny Him then the told time for justice has really drawn near. Except, that in any corner of the earth a people respond to Allah wholeheartedly due to which Allah's mercy must necessarily come to aid them over others. And then followed by a time for blessing to make them inheritors over earth for another age. But if global community desists, the consequences are worrying. These are the types of results which the chief of rebellion and his troops among jinns seek from the cosmic network together with men Kabbalists who associate with such jinnis for relationships indicative of verse: *Quran 72:6 "And there were men from mankind who sought refuge in men from the jinn, so they (only) increased them in (evil) burden."*

Doubtless many recipients of the Arabic Quran have lost their minds altogether to many slimy mines. Striking as it is the Quran is not a defender of Muslims and neither are Muslims stakeholders with Allah. Only those who behave justly and obey God alone will be successful. As faith is not a verbal claim nor a prestigious tag that people wear with beards and skull caps. Instead right actions alone are valued with God. In fact, it is a people who practice uprightness and can serve good; who chance to be successful. Be they among Atheistic naturalists, if they could believe; who deserve first allegiance to God more than a people who fell corrupt whom God detests. It could be a Westerner's flag in coming times for Allah—the Lord of the Worlds. The good men and women amongst you like your predecessors in the bygone days can in this new age once again be friends with God—the Lord of the Magnificent Throne.

History is now seen clearer unlike before as it repeats in cycles of the good, the bad and the worst. The Mongols had it and the Turks too; it is, as if troubled times will sieve new Romeos to God. For God had already blessed you when once your progenitors followed Jesus Christ the son of Mary. Remember their humble numbers as they grappled fists seeing their Master hunted down. Then God did spare Christ and the sincere ones in faith. And look they became the leaders for men to enter faith and a mighty revolution was instigated. Then there was Rome and the Christian compassion. It is easier said than done. If I should reach light; that belongs to me, like it belongs to you, I must accede to ideas that are true. To love your Lord above all assort yourselves and beseech faith. He expects that we partake fearlessly advocating truth. Rise to shoulder the Quran. The final message from God then you'll see your nation will have yet another life span. If not from under earth your vast landscapes will surely be transformed from above.

7.6 EXCLUSIVE REALITY

Heartbeat stars, discovered in large numbers by NASA's Kepler space telescope, are binary stars i.e. systems of two stars orbiting each other.[181] There is a bit of weird science for why they are named as heartbeats. It is said that their brightness over time maps out like a graph of the electrical activity of the heart like in an electrocardiogram.

Excerpts from Kepler and K2 published 22 Oct 2016 on nasa.gov: "...Scientists are interested in them because they are binary systems in elongated elliptical orbits."

"You can think of those stars as bells, and once every orbital revolution, when the stars reach their closest approach, it's as if they hit each other with a hammer," says Avi Shporer, NASA Sagan postdoctoral fellow at NASA's Jet Propulsion Laboratory, Pasadena, California, and lead author of a recent study on heartbeat stars. "One or both stars vibrate throughout their orbits, and when they get nearer to each other, it's as though they are ringing very loudly." "...Kepler, now in its K2 Mission, discovered large numbers of heartbeat stars just in the last several years. A 2011 study discussed a star called KOI-54 that shows an increase in brightness every 41.8 days. In 2012, a subsequent study characterized 17 additional objects in the Kepler data and dubbed them "heartbeat stars." Further data and research is required to characterize these unique systems.

Per the Quran's narrative discussed in chapter 5 and in sections 7.1 through 7.5, it is made clear that stars in the solar neighbourhood are designed for definite outcomes of security over trespasser jinns or serve other specific purposes. Various communications from Allah reach His Angels in the sky of this world like the revelations of divine information وحى (wahi) and matters concerning measure and destiny of things قدر (qhadar) and plans concerning busting of rebellion related happenings in this world. The key words حفظا (hifzuann) meaning protection and رجوما (rujumann) meaning termination by lapidation as seen earlier in section 7.3 speak aloud. In this chapter we cover heartbeat stars and their effectiveness in racing up to blockading the mammoth industrious actions of inquirer jinns.

The surah named at-takweer translated as -the wrapping of star-light- is found to provide Cosmic elucidation of heartbeat stars. The talk here is of a collapsed

---

[181] Heartbeat Stars' Unlocked in New Study on nasa.gov; Oct.22, 2016.

star besotting into a hide of its neighbouring star after having it yielded to duties and then receding.

> Quran 81:15-16 I (Allah) swear by the recede (of stars). Those that fade into neighbour (2nd of the twin star).[182]

In the verses that immediately follow this Cosmic detail attention is quickly brought to show the new build of context regarding revelations of the Quran and the importance of Gabriel providing teaching. An imprint duplicated also in surah named an-najam meaning the star.

> 17-18. And by the night as it departs. And by the morning as it dawns.

> 19-21. (That) indeed, this (Quran) is a word (conveyed by Gabriel) a noble messenger. (Who is) possessed of power with the Owner of the Throne, secure (in position). Obeyed (by other Angels) and trustworthy.

> 22-24. And your companion (Muhammad) is not (at all) hallucinating. And indeed, he sighted (Gabriel) in the clear horizon. And he is not a withholder of (knowledge of) the unseen.

> 25. And this Quran is not the word of a devil, expelled.

> 26-29. So where are you going? It is not except a reminder to the worlds. For whoever wills among you to take a right course. And not as you desire; but, as Allah wills- Lord of the worlds.

Coming back to verses: *Quran 81:15-16 "I (Allah) swear by the recede (of stars). Those that fade into neighbour (star)"*, here the Arabic phrase الجوارى الكنس (al-jawūar al-kunnas) is used. Winnowing this phrase to its root elements gives the meaning -the neighbour, the maid-. The neighbouring maid is also connoted to mean -a sweeper-. The etymology of "neighbour" connotes the behaviour of advancing and retiring or coming and going. The full phrase in non-Cosmic etymology connotes a wild bull or an antelope entering its hide, a place amidst the trees and in astronomical etymology it means stars hiding; thus, it is here

---

[182] Or another in the multiple star system.

interpreted to mean the stars disappearing into their neighbourhood hide.[183] This phenomenon of fading of stars into hides is further explained in surah an-najam, the star, by the example of Sirius B besotting into the hide of Sirius A after its collapse.

The role of heartbeat stars in security in space is explained as follows. In the solar neighbourhood when an invasive jinn infrastructure develops for grand scale eavesdropping; a nearby traversing Heartbeat is made to shift tracks and reorient its orbital vector gallop towards the heavily invested intrusion site. The star's elliptical orbit stretches to overrun the targeted investments. In case the investments are not heavy still deserving attention because of violations (or a potential build), a barrage of space objects from star zones suffices. Thus, the Jinn infrastructure is targeted to be destroyed by the approaching heartbeat. At this point it is deduced that one of the two binary stars undergoes an implosion when needed.[184] The Angels implode the main sequence star on orders from High commission casting tremendous radiation energy destroying the much efforted jinn establishments. Therefore, to secure lines of wires coming from high constituency of Allah. Heartbeat stars operate on demand on fast cruises by reorienting to sweep radiation and also to overrun & implode causing destruction in the pockets of the first Cosmic sky of this world.

Chapter 53 titled An-Najam (The Star) in the Quran is centred around the Sirius star system indicative of verse 53:49. The chapter is found to dispel associated myths [185] surrounding this popular star that Allah Almighty is the only benefactor, that it is He who is the Lord of Sirius and nothing else but Him remains the effecter of what He intends on earth and in the Cosmos. This is in

---

[183] In Lane's Arabic lexicon pages 5628 and 6008.

[184] An implosion is simply the opposite of an explosion. In an explosion, matter and energy fly outward, but in an implosion, matter and energy collapse inward. All implosions will need some sort of pressure from the outside pushing-in to cause the object to collapse. Resulting in enormous energy outputs and receding into a concentrated core unlike full scale disintegration.

[185] Ancient Egyptians noted that Sirius rose just before the sun each year immediately prior to the annual flooding of the Nile River. Although the floods could bring destruction, they also brought new soil and new life. Fittingly, Osiris, whom Sirius may have represented, was a god of life, death, fertility and rebirth of plant life along the Nile.

stark contrast to other world cultures where the star Sirius is quite popular being attributed godly status and an ingredient to various superstition.

Quran 53:49 And Indeed that, He (Allah) is the Lord of **Sirius**.

Reading the first set of verses from An-Najam, a background ushered with astounding reality is discovered vis-à-vis just how the matters related to us are drawn in from the Cosmos.

Quran 53:1 By the Star, at its collapse.

2-4. Your Companion (Muhammad) is neither astray nor misled. And he does not speak by whims. It is not but a revelation revealed.

5-8. Taught- by him (Gabriel) intense in strength. One of soundness, so he rose (to his stately physical form). Awhile he was in the upper (Cosmic) horizon. Then, he descended, as directed.

Verb based translation of verse 9:

9. There he modelled twin-stars further nearer.

Noun based translation of verse 9:

9. Then he was around twin-stars (or) further nearer.

10. From where he, revealed to His (Allah's) Servant (Muhammad) what was stipulated.

11-13. The (Prophet's) senses did not fault that which he saw. Will you then dispute with him over what he saw? And-Indeed he saw him in yet another descent.

In retrospection of verse 9 let's ascertain whether the verb based, or the noun-based sentence builds the correct interpretation. The rationale developed is as follows: the particle ف (fa) in Arabic grammar plays five different roles. It acts in the prefix as a particle of cause, to supplement, to result, a coordinating conjunction and a particle for resumption. In this section (Arabic: juz) verses 53:1-10 of the Quran discussed here- ف (fa) is in the capacity of a particle for resuming several astronomical sequences. After Satan had rejected honour of mankind; vulnerability had to be managed and thus preparations assumed leading to the backdrop of these verses. The particle ف (fa) weaves the Cosmic story from Gabriel's preparation to descend and then covers the event of a collapsed star elucidating it in verse 9 where ف (fa) occurs the second time after its first citation in verse 6 connecting events discussed in verses 53:7-8 to the celestial descent as a precursor to implosion of Sirius B. And then the resumption is again effective in verse 10 where scriptural revelations given to a series of prophets is predicated before the scripture is given to Prophet Muhammad by the use of the particle ف (fa) which determines the eons of human activity on earth. Let's see verse by verse what revelation this chapter holds.

It begins with a thematic introduction detailing a Cosmic event in its opening verse.

> 1. By the **Star,** at its collapse.

The next four verses draw the context to engage with the disputants of Prophethood. Gabriel is also introduced as the main performer in verse 5.

> 2-5. Your Companion (Muhammad) is neither astray nor misled. And he does not speak by whims. It is not but a revelation revealed. Taught—by him (Gabriel) intense in strength.

Next, the Cosmic context of verse 1 is found to resume midway in verse 6, where ف (fa) occurring for the first time prefixes the verb فاستوىٰ (fa-istawa) translated as 'so he rose'.

> 6. One of soundness, so he (Gabriel) rose (to his stately physical form).

As mentioned above, Gabriel rose to his stately disposition, preparing for a celestial dive. From here he launched into a descent towards another point as revealed in the following verse.

> 7-8. Awhile he was in the upper (Cosmic) horizon. Then, he descended, as directed.

So, what happens after this descent? This is answered in the next verse which is the fulcrum of this event. There are two possible interpretations for it, and both are discussed as follows:

<u>Verb</u> based translation of verse 9:

> 9. There he modelled twin-stars further nearer.

<u>Noun</u> based translation of verse 9:

> 9. Then he was around twin-stars or nearer.

Arabic transliteration:

> 9. Fa kana qhaba qhausaini **au adna**.

Let us verify the verb-based interpretation first. In verse 9 the Arabic word قاب (qhaba) picked for a verb has several etymologies which mean: to induce a change, to manifest a residual outcome like the cracking of an egg to reveal a fledgling, emptying of earth to dig a hollow in round form etc.[186] In the light of such verbal inferences, a case for a verbal sentence is established. What is more, the verse begins with the form كان (kana) translated here as "there he" which in fact means "he was", this further strengthens for an explicit verb interpretation. i.e. the form فكان (fa-kana) for the syntax of verb قاب (qhaba) is thusly interpreted as: *'there he collapsed a star and modelled the twin-stars in further nearer orbits.'*

To sum-up 53:6-9, Gabriel rose, then while he was there, he was directed to descend to the Sirius star system where he collapsed a star and modelled the twin stars into nearer orbits. This event is the pre-cursor or the background for the thematic opening verse where Allah swears a mighty oath referring to this collapsed star in verse 1: *Quran 53:1 "By the Star, at its collapse."*

Next, we analyse the noun-based interpretation. To start with, the verb فكان (fa-kana) makes an incorrect syntax for this interpretation because فكان (fa-kana) meaning "he was", is a perfect third person verb. A verbal phrase is incomplete if the verb is not complemented. Present noun-based interpretations of this verse rely to complementing this verb in contexts of the preceding and the following verses. However, to enforce a nominal syntax قاب قوسين (qhaba qhausaini) in the verse would translate as: -

1. Then he was distance two-bows or nearer.

---

[186] In Lane's Arabic lexicon pages 5474-5475.

2. Then he was around twin-stars or nearer.

For nouns, as opposed to English which has a singular form for one and plural form for more than one, Arabic has three forms viz, singular form, dual form & a form for three or more. In verse 9 the noun appearing in the Quran is قاب قوسين (qhaba qhausaini) where قوسين (qhausaini) is in its dual form. Lane's Arabic lexicon is found to explain only the singular form of this but what is interesting is that Lane provides both the non-cosmic and cosmic explanation for the **singular** قاب قوس (qhabu qhausinn) as follows:

1. Non-Cosmic -> between them is the measure of a bow.
2. Cosmic -> the distance between two stars.

Hence it is established in lexicon that technically both non-cosmic and cosmic interpretations are applicable. Now it needs to be proven that neither of these two fit the verse.

For brevity sake if one were to apply Lane's singular explanation to dual, as the Quran has the word in dual form, these two noun-based interpretations would translate as: -

1. Non-Cosmic context ->*Then he (Gabriel) was distance two bows (or) further nearer.* Which is absurd as It is bizarre to imagine an Angel in his real physical appearance to be in proximity before a man spaced at about two-bows length. It is not possible to see the elevation of a house at two bows length forgo a stately physical Angel of the make like Gabriel. Therefore, this nominal case stands omitted.
2. Cosmic context -> *Then he (Gabriel) was around twin-stars or nearer.* This is also incorrect as according to the Lexicon, the singular itself brings out the meaning as 'distance between two stars.' And it would also render قاب (qhaba) as a noun to be dysfunctional to be used with a dual form قوسين (qhausaini). Thus, the Quran's usage of dual form with the verse ending interjection implying '**or nearer**' is clearly intending something entirely different unless we want to further muddy the interpretation with say Gabriel hung some distance between binary star system and earth.

Some online language translators also translate the dual form قاب قوسين (qhaba qhausaini) as "around the corner" which is also omitted for the same reason as the non-cosmic version stated here. And if the corner is asserted as alluding to around Sirius in a symbolic way in celestial sense it still stands omitted for the same reason of ambiguity of the verse ending interjection أو أدنى (**au adna**) implying '**or nearer**'.

Thus, it is evident that all the noun-based interpretations for verse 9 are erroneous and the most suited interpretation is the verb-based interpretation which is:

'There he modelled twin-stars further nearer' which in the context of verse 1 is understood as *'there he collapsed a star and modelled the twin-stars in further nearer orbits'*.

Further, as Quranic evidence is found to lay explicit emphasis on unspoken mystery phenomenon unfolding before Prophet Muhammad as shall be seen in verses 53:10-18 (discussed ahead) and openly asserts him sighting the Angel plainly in his stately physical form and as this data is well supported in the biography (Arabic: seerah) of the Prophet too that Angel Gabriel had his wings filling the Cosmic horizon and had become permeable for him to see, therefore, from above analysis it eventualises for verse 53:9 to assume a context in the celestial space in its verb form and not on earth at a distance of two bows, in the Cosmic surah, the Star.

Now rationalizing this verb-based rendering, it consists of a verb descriptive of the status of a subject or a doer followed by which predicates complementing part for this verb by an action followed by the object on which the action is done. The first verb, فكان (fa-kana) translated as "there he was" is acted by the subject "Gabriel" upon the object (a noun) قوسين (qhausaini) the "twin stars" and the action performed is the second verb قاب (qhaba) which is Gabriel's act of modelling of the twin star system. This is the implosion of Sirius B main sequence star into a white dwarf resulting in prominence of Sirius A as an emerging brilliant bright star, referenced by Lane's lexicon in the etymological sense as: Cracking an egg open to reveal a fledgling or such as "disclosing an affair" (addition) per Lane's lexicon. Therefore, the substantiated interpretation would read as— **"There he modelled twin-stars further nearer"**, indicative of spacing orbits between the stars to have gotten closer after Sirius B imploded from main sequence into a white dwarf. Resulting the fast gallops of Sirius system to rest in closer orbits. This is the exclusive reality found in the Quran to those who harken to its message. After the implosion of Sirius B, the disbursed star material contributed to the prominence of Sirius A.

Since then Gabriel has been responsible for delivering revelations to Prophets and Messengers in the line of Adam up to Noah, and in Noah's line, and to others from those who boarded the Ark with him, and in their line who spread to various regions of the earth siring nations and then to Abraham and in his line. Thence, the prophecy was isolated to those who sprung from Abraham finishing with Muhammad. Gabriel did reign in this business in the nether world. The cosmic event of Sirius B's implosion can therefore be dated to have occurred when Adam's resettlement from the garden in earth was concluded by Allah's high assembly as knowledge thereof was given in favour of bringing mankind to the fore of a grand stage of finding their purpose on earth.[187] Another probable time could be since around Noah's time, at when full-scaled revelations from Allah had assumed as until up to then no god besides Allah was worshipped by any people in preceding generations.[188] And therefore many determinisms to punish mankind on earth weren't necessary. However as noted earlier, man is the late arrival to planet earth. Prior to emergence of mankind, earth was long filled with its various creatures of flora & fauna along with jinns and various species of their kind. Cosmic barricading for jinns began with an implosion of Sirius B. Since then, the Jinn are gradually quarantined and under constant reconnaissance for their ability to tap divine information concerning human affairs. As nations grew and our affairs had to become complex Allah has despatched additional stern infantry of Angels in excess to engineering several security mechanisms in the first Cosmic sky such as the Sirius implosion to regulate activities of Jinn and mankind. It is also theorized that the implosion was perhaps prior to the mission

---

[187] Quran 2:38 We (Allah) said: "Step out all from here (and work your bread); and if as is sure there comes to you Guidance from me, whosoever follows My guidance, on them shall be no fear nor shall they grieve.

[188] Quran 10:18-19 And they worship other than Allah that which neither harms them nor benefits them, and they say, "These are our intercessors with Allah "Say, "Do you inform Allah of something He does not know in the heavens or on the earth?" Exalted is He and high above what they associate with Him. **And mankind was not but one community** (united in religion), but (then) they differed. And if not for a word that preceded from your Lord, it would have been judged between them (immediately) concerning that over which they differ.

of Jesus Christ as seen in section 5.3. Regardless of its exact date of happening, it is clear that many additional celestial rearrangements within solar system were exacted as seen in section 7.2, before the final deliverances of divine document to Prophet Muhammad in Makkah in fulfilment of the prophecies of the old such as Genesis 17:20 & 21:18 found in the Bible.[189] Foretelling, that Ishmael will be a great nation like Isaac which means even in his line prophecy will become awarded. With these milestone changes to regulate their influence, jinns could perceive that their Cosmic explorations weren't the same as before. As jinn affairs are markedly different than ours, vulnerable human needs have thus been meticulously taken care of by Allah's exclusive mercy: *Quran 55:35-36 "Upon you be sent space objects of a fire brand and of smoke. And you won't be defended. So, which of the favours of your Lord would you deny?"*

Aged forty in 610 A.D before prophethood, Muhammad would retreat privately into the cave of Hira in mount Nur, also known as mountain of light. Here he would often betake himself in search of truth, away from the vanities of his people. It was on one such occasion while he was on his foot to return home that in the horizon he was meted by a cataclysmic event. Gabriel showed up in his true stately physical form. From an altitude on mount Nur in the outskirts of Makkah, at every turn of himself Muhammad could see the sky filling with the Angel. Allah had chosen Muhammad as his messenger- a modest man from a noble family, well mannered, well known for his truthfulness and honesty, who deeply cared about the truth of life and having the disposition to see the right from wrong. On that fateful day on the mountain Muhammad thus, was greeted by Gabriel as the chosen Prophet of Allah and the very first revelation was delivered to him which is arranged in the Quran from verses 96:1 to 5.[190] Thus introducing Lord's Kingdom in a grand way to a people who were severed from Prophecy: *Quran 36:6 "That you may (O Muhammad) warn a people whose forefathers were not*

---

[189] Genesis 17:20 And as for Ishmael, I have heard thee: Behold, I have blessed him, and will make him fruitful, and will multiply him exceedingly; twelve princes shall he beget, and I will make him a great nation. And Genesis 21:18 Arise, lift-up the lad, and hold him in thine hand; for I will make him a great nation. (KJV)
[190] Quran 96:1-5 Proclaim! (or read!) in the name of your Lord and Cherisher, Who created—Created man, out of a bearing (on the Uterine). Proclaim! And your Lord is Most Bountiful. He Who taught (the use of) the pen. Taught man which he knew not.

warned, so they are unaware." Overwhelmed at the turn of events and having seen the unseeable, he rushed downhill for cover and asked his wife Khadijah to shroud him as things beyond imagination had befallen him. Later as revelations assumed regularity, Gabriel's deliverances from the unseen dimensions became normative. Gabriel appeared to Muhammad in his true stately form only twice. On other occasions he appeared in human form, as is well known in traditions (seerah of the Prophet) and a concept supported by the Quran. Most part of the revelation were transmitted as information only and rarely any messages were brought in by appearances of walking visits, all becoming part of the book.

> Quran 53:10-12 Then he revealed to His Servant (Muhammad) what was stipulated. The (Prophet's) senses did not fault that which he saw. Will you then dispute with him over what he saw?

The ف (fa) found in beginning of verse 10, is a particle for resumption of a second theme, it is the revelation of Quran itself. Verse 53:2-5 with verse 10 are clarifying statements of resuming this revelation of Almighty Allah, in the context of prophecy revered in traditions of Judaism. Here, it is implied that Gabriel assumed his deliverances to Prophet Muhammad after the pause which had been in effect since the mission of Jesus Christ to children of Israel. Expectants of prophecy among Jewish faith today must know that an abrupt closure of prophecy for 2000 years now is not indicative of its resumption in future after a line of Prophets had already successively shown up amidst them. And when they rejected Jesus the Maseeh (Christ), the office of Prophethood after that was closed with Muhammad.

Verses 53:11-18 thereafter reveal some prominent Cosmic events exclusive to Prophet Muhammad's sighting of them. Watching Gabriel plainly in the horizon at first revelation was clearly the first of the great signs. In his second sighting of Gabriel, Prophet Muhammad had experienced time-travel. Since Albert Einstein's postulated theories on time travel, we have been anticipating an invention of a futuristic time-machine. Also achieving speeds nearing c remains at large a distant dream. Then still outside our reach is teleportation. Quantum physics shoulders some burdens on cracking the idea of time travel via teleportation on earth and in space. With science of quantum entanglement already demonstrating teleportation of atomic and photonic states,[191] the future may promise some unpredictable science breakthroughs on this front. If we could head science in the direction of propulsion of states of quanta using earthly skies i.e. Magnetospheric & Plasmaspheric for translocation then perhaps demonstrations achieved by the knowledgeable men of Prophet Solomon could be demonstrable in our science endeavours too.

> Quran 27:39-40 Said an 'Ifrit, of the Jinns (to Solomon): "I will bring it (the throne of Queen of Saba) to you before you rise from your council: indeed, I have full strength for the purpose, and may be trusted." (Then) Said one of men (in Solomon's council) who had knowledge of the Book: "I will bring it to you within the twinkling of an eye!" Then when (Solomon) saw it placed firmly before him, he said: "This is by the Grace of my Lord! - to test me whether I am grateful or ungrateful! and if any is grateful, truly his gratitude is (gain) for his own soul; but if any is ungrateful, truly my Lord is free of all Needs, Supreme in Honour!"

The surah Al-Isra in the Quran is named after space time-travel. The opening verse carries the Arabic word أسرى (asra) translated as **take** or '**took**' bears more meaning into it at a fundamental level. It portrays the mysterious night journey of Prophet Muhammad from al-Haram, Kaba in Makkah to al-Aqsa holy temple in Jerusalem: *Quran 17:1 "Glory to (Allah) Who did **take** His servant for a Journey by night from the Sacred Mosque to the farthest Mosque, whose precincts We did bless-in order that We might show him some of Our Signs: for He (Allah) is the One Who hears and sees (all things)."* The word أسرى (asra) means captivity. This word elegantly alludes to the entanglement of Prophet Muhammad for space time-travel in scientific connotation. Allah had the Prophet take this one-of-a-kind and incredible night Journey in a demonstration of his might to captivate his human heart in awe and wonder. He had the Prophet night journeyed to grow him in certainty of Allah's magnificence. Prophet Muhammad was lifted to traverse a course in the cosmos and then was returned to Mecca in the same night. Such a journey requests science terms be called-in to explain the anomaly. He was taken via space time-travel, first to the holy land of Jerusalem. Thereafter

---

[191] Title search: Big step for quantum teleportation won't bring us any closer to Star Trek. Here's why; on sciencemag.org; by Adrian Cho Sep. 19, 2016.

the verses reveal him to have visited the planet of eternity wherein at the Lote tree is referenced to exist; which marks the beginning of the enduring gardens of Paradise.

> Quran 53:13-18 And indeed he saw him in yet another descent. At the prohibited **Lote-Tree**. Where there is—the enduring gardens. As the Lote Tree was covered by that which covered (it).[192] The sight (of the Prophet) did not swerve, nor did it discomfit. Indeed, he saw from his Lord, the greatest signs.

European nettle tree, also known as European Hackberry or Lote tree (Celtis Australis) is a deciduous tree. It's leaves as well as the small fruit seems to have proven medicinal properties.[193] Its astringent leaves and fruit are found to treat amenorrhoea, heavy menstrual & intermenstrual bleeding and colic. Attention is drawn here to the fact that it was the prohibited tree-type from which Adam and Eve ate when asked to dwell in the garden in earth. Consequently, their nakedness had become evident to them which brought them to face procreation and many determinisms by exercising of human freewill in a test laced before us prior to dying. The Arabic word حرام (haraam) is translated as prohibited. Similarly meant is the word ممنوع (mamnu) formed from its trilateral root منع (mana) which means to forbid. Both, of these words did not appear in the story of Adam and Eve in the Quran in relation to eating from the tree. Though the command to not eat from the tree came with consequences: *Quran 2:35 "And We said, O Adam,*

---

[192] Reported by Asma daughter of Abu Bakr (early companion) in Mishkat-al-Masabih translated by James Robson, vol. II p. 1201 (Bk. XXVI Fitan). Also, from Ahmad ibn Hanbal's collection reported by Abdullah bin Mas'ud (early companion) that it was golden butterflies like Danaus genutia also called common tiger that had covered the Lote Tree. A common fact akin to monarch butterflies migrating toward south of America yielding to warmer weather patterns from severe winter in North America, see **Figure 33**. It is a bit of mystery as to how fourth-generation butterflies fly back to the same orchards to shroud them. However, this narration here is a plausible citation because of seventh century unawareness of butterflies shrouding trees; knowledge of this phenomenon wasn't in common awareness. That too when it is coming from across continents of American lands when possibly it wasn't a known thing among the eastern civilisations fourteen centuries ago. Even in modern times many but the natural enthusiasts aren't aware of this behaviour of butterflies to shroud trees. Therefore, for this information to have had a narrative substantive of elucidating a Quranic verse its authenticity is thus determined.

[193] Online for more info. Also see pfaf.org.

dwell, you and your wife, in garden and eat therefrom in (ease and) abundance from wherever you will. But do not approach this tree, lest you be among the wrongdoers." But a long while after Adam had relished in the gardens, the chieftain jinn, casted whispers of delusion in them to eat from it. As a result, they ventured into the forbidden and the scripture records that unlike Satan, they were conscious of their error and turned to Allah in repentance. Allah, then forgiving of their disobedience, brought them before what was long destined. By their own precipitated act, they graduated to bear responsibility for freewill. In verse 53:14 here **"at the prohibited Lote-Tree"** the word المنتهى (al-muntaha) does not mean 'prohibited' in the sense of the English word either. Instead the word المنتهى (al-muntaha) is interpreted as a boundary to enter the everlasting gardens in relation to the humongous Lote-tree proportional to the sizes fitting of Paradise planet. Therefore المنتهى (al-muntaha) is the finality of a symbolic boundary which if attained the end goals come into effect. The Lote-orchards, I believe en route to Paradise is to rejuvenate physical & sexual wellbeing in folks released from the pits of hellfire post their suffering penalties. As, even a little faith in Allah is classified as worthy for an entry into the gardens of eternity post their cleansing in the pits of hellfire. In this grandest of journeys and the majestic horizon, Prophet Muhammad was shown among the greatest of signs, the might of God to undertake Muhammad on a tour in fragments of time, the planet wherein are the splendid gates to the heaven and also is the fiery pit. Also, in that Cosmic Skyline was the descent of Gabriel when Muhammad sighted the arch-Angel for the second time.

Quran 53:13 And indeed he saw him in yet **another descent**.

What is interesting to note here is the Quran drawing attention to Gabriel's 'descent' indicating that unlike the Prophet who was teleported, the Angel travelled to the said space. It could be inferred here that teleportation is not their usual mode of travel which is in-line with the theory of Angels speeds proposed in section 7.5. Therefore, Gabriel had to travel to where Muhammad was teleported to, and he arrived, revealing his stately physical form for the second time to him. As the Prophet had efforted much to win the hearts of prospective believers by his selfless sacrifices and faced great harshness and hostility from the residents of Makkah in his early years, Allah thus, showed him the wonders of the cosmos which He deemed fit for his Prophet to see. Just about when things were worsening in Makkah gripping the new converts with maximum sufferings and about when the Prophet had lost his two biggest aids, his wife Khadijah and uncle Abu-Taulib (chief of his tribe) who shielded him from the vengeance of disbelievers. Allah willed for him the unbelievable and disclosed to him matters in this miraculous night journey which none alive is permitted to see. It was a fitting privilege for the one designated as the final Messenger of God so that an argument of Muhammad receipting divine communication and his mysterious space time-travel is projected upon the rest of mankind till doomsday.

In seventh century, Arabic was largely a spoken language and there weren't any standard literary works transcribed yet. As the rudimentary Arabic syllables took an evolution thereafter to become an advanced written script, it captured the breaths of all spoken sounds in their tongue.[194] The Prophet was unlearned, and he couldn't transcribe any book. The Quran witnesses this and so did his people, who knew him and his life well: *Quran 10:16 "Say (O Muhammad), 'If Allah had so willed, I would not have recited it to you, nor would He have made it known to you, for I sojourned with you a lifetime before it (Quran). Then will you not reason?'"* Gabriel continued to teach him to recite and soon the Prophet caught on to recite well with him.

> Quran 29:48-49 And you were not reciting before it any book, nor were you (able) to inscribe one with your right hand. In that case, the doubters had a chance to falsify.
>
> Rather, it is distinct verses (soaking) the conscious minds of those knowledgeable. And none rant against verses of Allah except the wrongdoers.

Summarising the narrative in this section, Angel Gabriel at the rising of humanity to face their test for faith and conduct, launched himself in a dive from the upper celestial horizon within Milky Way and assumed a near distance toward earth; his assumed trajectory is traced for a plausible departure from the Orion nebula descriptive of بالأفق الأعلى (bil-ufqhil-āla) in verse 53:7 to the Sirius binary system (see section 6.9). This dive per the approximation of 1 second per light year (section 7.5), Gabriel's journey from the Orion Nebula to Sirius would have

---

[194] The early Quran manuscripts bear evidence of this fact. Awhile the Quran was written, the script also evolved into more expressive phonics and artfully cursive and lucid in calligraphy making it easy for the non-Arabic world to read.

lasted 1300 seconds or 21.6 minutes. The words الملإ الأعلى (al-mala-il-āla) translated **high (exalted) assembly** in this context of verses 38:69-70 are the legions of Angels arrayed in the Orion nebula where Gabriel held his seat after God's creation of earth and life in it. Concerning the decisions that are drafted from there for various work orders on earth, Prophet Muhammad was made to say: *Quran 38:69-70 "It is not up to me to have knowledge of the **exalted assembly** (of angels) when they conference (with God). It is only revealed to me (to declare) that I am a clear warner."*

When since Gabriel was despatched from the upper celestial regions 'بالأفق الأعلى (bil-ufqhil-āla)' to the near-earth gallop post at the star Sirius, he has since then grandly reigned in this office, being instructed by Allah with knowledge to manage vaults concerning Allah's revelations to mankind. And has since time and again revealed scriptures from Allah to various nations and in the past in many vernacular languages. For this reason of divine deliverance and earthly affair management, Gabriel first arrived at Sirius- the brightest and among the closest stars to earth. His actions at Sirius system, descriptive of the Arabic phrase قاب قوسين (qhaba qhausaini) unfortunately are mistranslated thus far in the traditional Quranic corpus. Gabriel's actions interpreted in this work as having modelled the binary star system, re-orienting its orbital vector towards earth and understood to mean the implosion of Sirius B main sequence star into a white dwarf, is a detail conveniently found conveyed- a miracle to those beholders of modern science. When, the Quran is found emphasising the collapse of a star and arch Angel's cosmic plunge then traditional meanings for the Arabic word قوسين (qhausaini) translated as two-bows is clearly unyielding and wrong. And قوسين (qhausaini) deserves a connotation from an Astronomical sense of etymology. In fact, blunt translations of two-bows are interpretations sprouted due to external influences. The story concerning first revelation in the cave of Hira on mount Nur has become subject to spurious narratives in certain bits. It is alleged that since Muhammad was unlearned the Angel embraced him and constricted him a few times before he was relieved to follow words of revelation after Gabriel. The stories also report that until then, he could not follow recitation but kept saying, I am unlearned. It is true that first revelation pressed him hard and demanded him to recite or read as the Quran announced in Arabic اقرأ باسم ربك (iqhra bi-ismi rabbik) meaning 'Read! In the name of your Lord'. But the zeal, apparently typical of Arabic tribesmen's prose, is carried further. It is common for story narrators to add or exaggerate instances which was understood for prose. Wrongdoing isn't implied where it brings out a natural way of expression. To exaggerate storylines and to impart expressive uniqueness demonstrates talents and skill of the speakers. Therefore, Hadees (or Hadith) narrations cannot be verbatim concluded to be having all elements true in its entirety of communicated narration. Abu Bakar the first adult male Muslim convert and a long-standing companion in Prophet's mission and who later became the first Khalifa of new Muslim community expressed much hesitation to quote the Prophet verbatim. He is found to have added disclaimers in whatever he narrated on behalf of the Prophet, something like 'similar was the Prophet's expression' and so on. Unfortunately, late comers to Prophet's mission have indiscriminately scored maximums for transmissions of narrations on behalf of the Prophet. For example, in a case of Prophet's expression that he saw Gabriel full of glow and aura (Arabic: Nur), it is easy to deduce, that it could be said of Gabriel as made from light. Which is merely an interpretation for its first information. Not realising that eyesight is facilitated by light, it was easy for transmitters to communicate that Gabriel is made from light. However, as the Quranic revelations upon Prophet Muhammad became a recital recited in prayers in the Prophet's company, lengthier than our present standards of making prayers, it made it possible for its verification in vocal gatherings for its correct deliverances. The Quran was committed to memory by its earliest followers, which was later codified in a single book (Arabic: Mushaf) like a print to copy format. There is one straight answer to all criticisms arguing about the Quran's preservation of content—The Quran was revealed and still is recited in a dialect of its revelation upon the Prophets tribe, the Quraysh in Makkah which is unambiguously available in intact version without a second. Sensing not the need to further clarify, the Quran is exquisite in detailing that, Angel Gabriel made a stately appearance; filling the horizon to Muhammad which did terrify him. The translations, therefore, that he came close to Muhammad at about two-bows length is inconsistent and falls within the hearsay story which concludes that Gabriel had embraced him and further constricted him. Constricting is clearly not at two-bows length anyway but squarely zero spaced! However, spurious narrations surrounding the personality of Prophet do not end here. Like this, obvious multiple inconsistent narratives of several other incidents have been reported making it very vague and difficult to confirm the actual verbatim of the Prophet in most narrated content.

Thus, it is inferred that Gabriel wasn't tickling his lips by his Cosmic descent directed toward Sirius, but he caused an implosion of Sirius B as reckoned in the opening verse of chapter 53. Gabriel's activity at Sirius soon after his reaching the binary-star system, when both stars were in their main sequence age, was to implode the major of the two. In the process also destroying the outposts of jinn space stations in the interstellar sky en route to earth—in the first Cosmic sky. As Sirius made a prominent star system close to earth among few other stars in the neighbourhood, jinns were already gearing up efforts to scale and cover zones by their foresight of Sirius as the Angels most likely canter post least expecting its implosion. When Adam with Eve relished their pre-examination era, chieftain Satan had leaps of efforts to learn first information about Allah's dealings with Adam and his impending progeny. Therefore, the goldilocks zones of Sirius among other near star systems were most likely flooded with jinn instrumentation. The burst of energy and radiation from the main sequence Sirius B casted an inferno that crumbled illegal activity in the interstellar regions that encompassed confidential space en route to earth. This also explains the trajectory of Sirius's orientation vector towards solar system and its progress known via astrometric measurements to account for 5.5 km per second towards our sun.

With respect to the Sirius star system without clinching what is known in modern Astronomy; the interpretation of Quranic text here is derived aided by modern understanding regardless of what is making present science. Tell me if this is not a miracle given by God to serve for lasting days of human scientific inquiry into the unknowns of what surrounds us! Alarmed, naturalist Astronomers perhaps may criticise this narrative; concerning the inferences given to the timing of Sirius's implosion. In that case they shoulder the burden of reproof. However, this study is from my limited lenses of knowledge. Meaningfully, I'm ever ready to concede to best evidences produced.

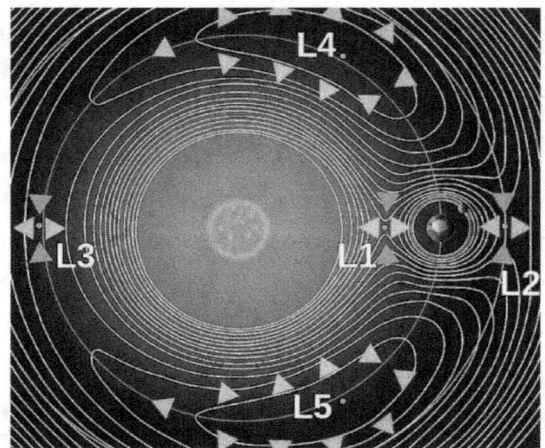

Figure 23 *L-points/Lagrange points.*
Lagrange points located around in sun-earth system; GAIA satellite observatory sits in L2.
Image sourced: Wikimedia commons; Credits: NASA.

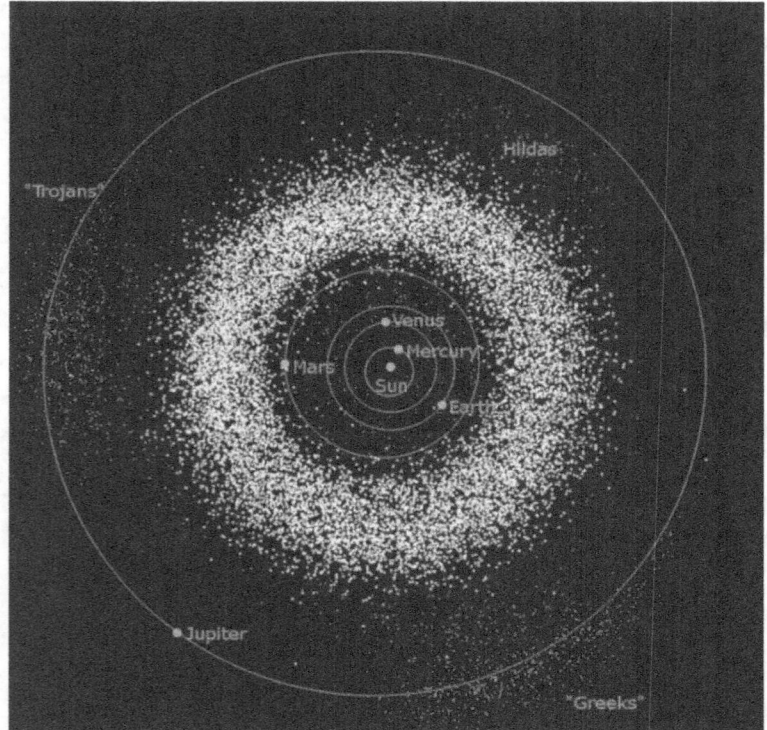

Figure 24 *Asteroids belt, Trojans, Hildas, and, Geeks.*
Credits: Mdf at English Wikipedia.

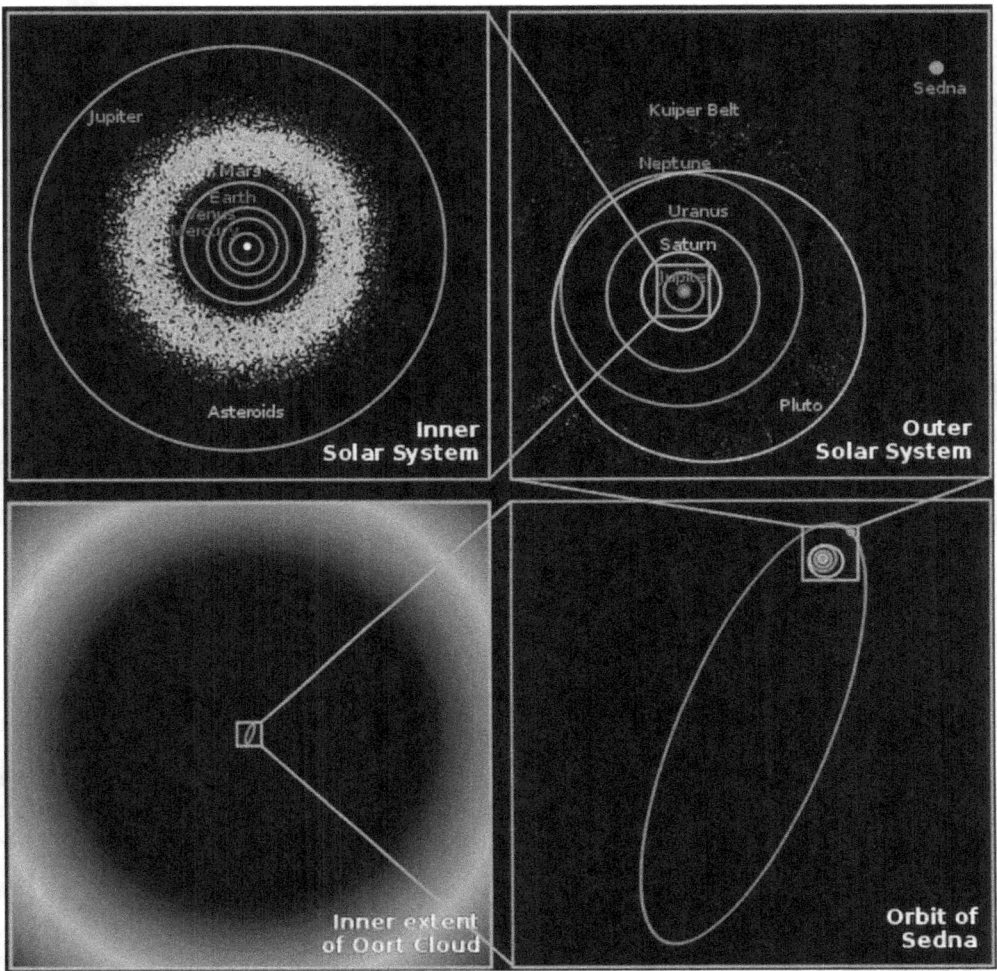

Figure 25 *Oort cloud to Asteroid belt.*

Presumed distance of Oort cloud; Mysterious Sedna orbit; Kuiper belt; Asteroid belt. Image sourced: Wikimedia commons; Credits: NASA / JPL-Caltech / R. Hurt.

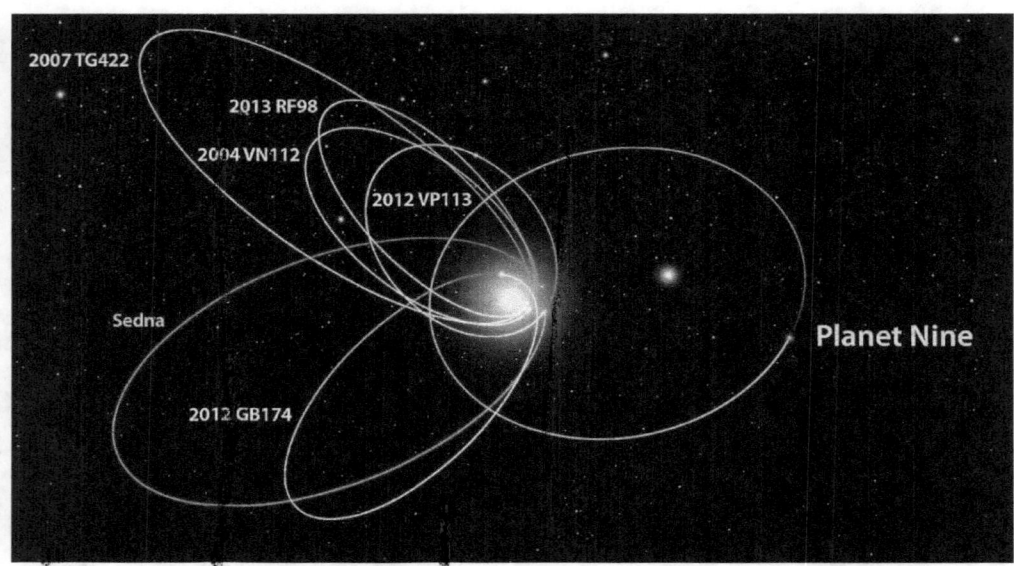

Figure 26 *Mysterious orbits of many distant bodies beyond Neptune.*
Image credits: NASA.

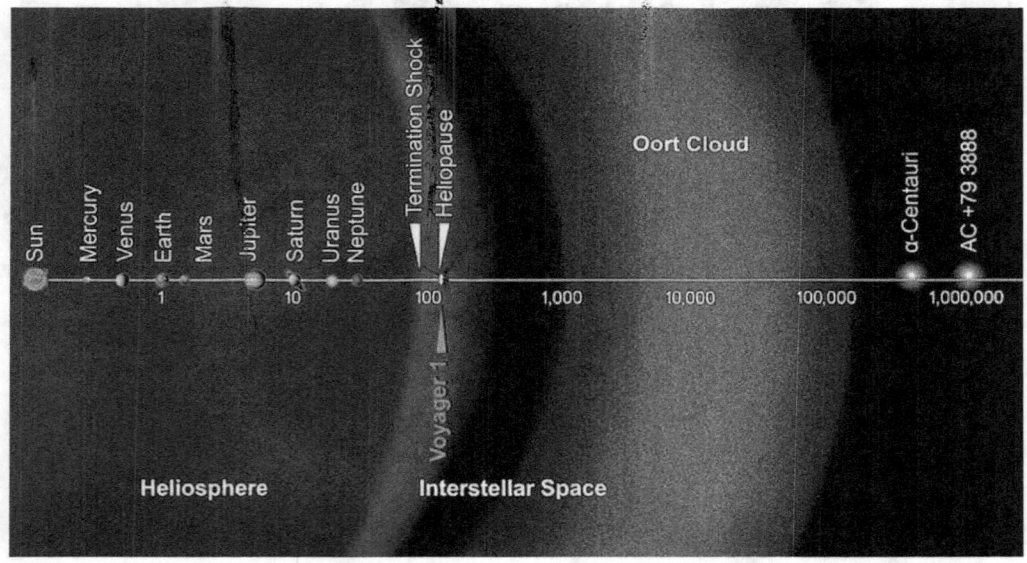

Figure 27 *Voyager 1 in relation to Oort cloud.*
Credits: NASA / JPL-Caltech; Image sourced: Wikimedia commons.

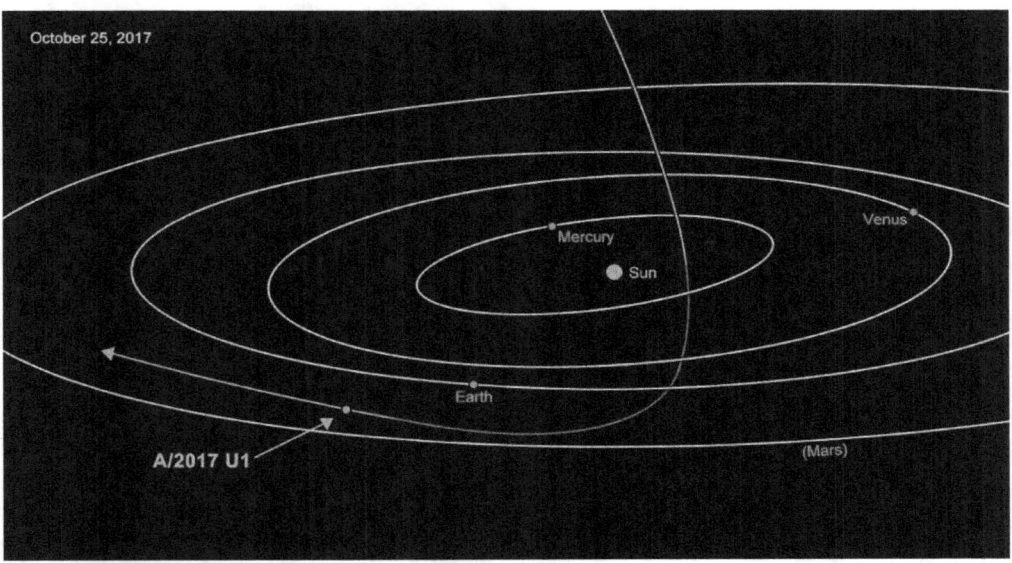

Figure 28 *A Comet from another Star- Interstellar missile.*

A/2017 U1 is most likely of interstellar origin. Approaching from above, it was closest to the Sun on Sept. 9, 2017. Traveling at 27 miles per second (44 km per second), the comet is headed away from the Earth and Sun on its way out of the solar system. Credits: NASA/JPL-Caltech.

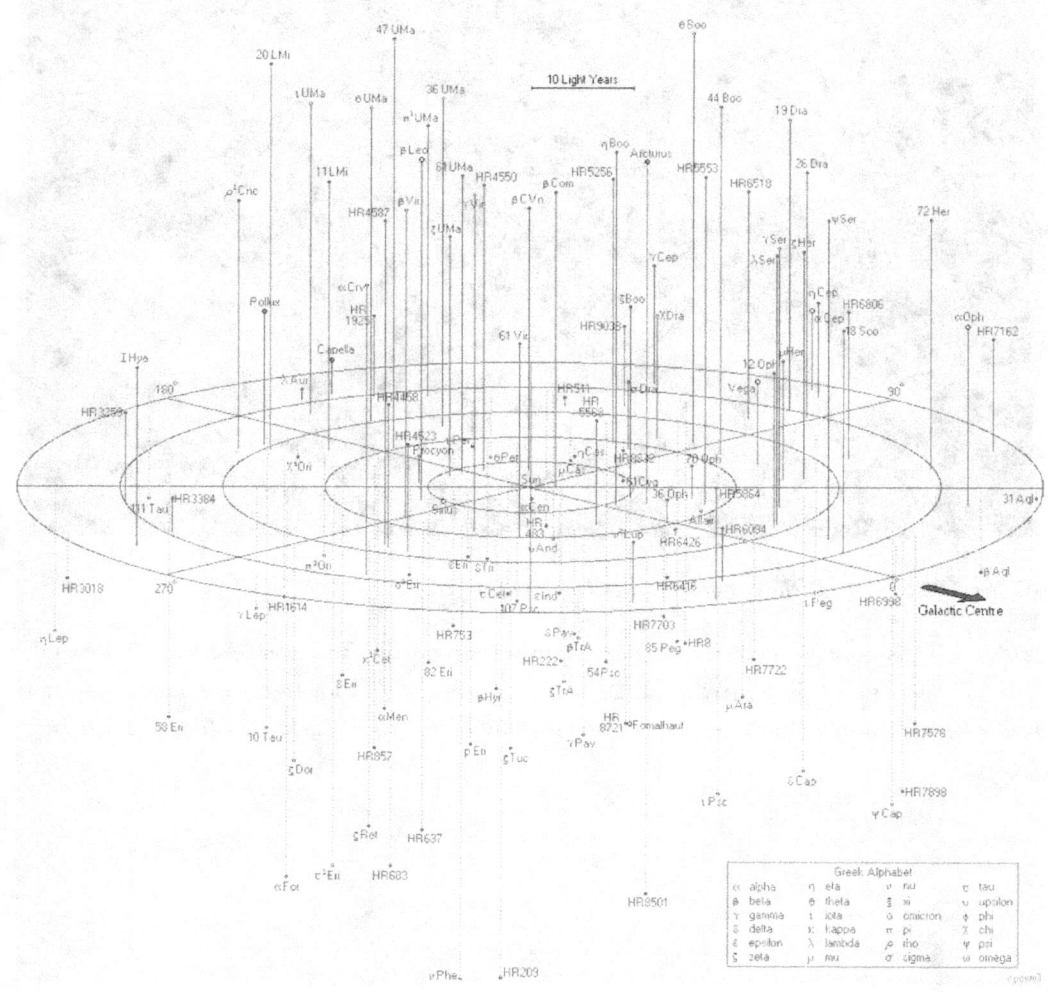

Figure 29 *Solar neighbourhood- Within a volume of 50 ly.*

There are 133 stars marked on this map. Every star brighter than magnitude 6.5 within 50 light years, seen with the naked eye. Map shows the brightest 10% only of all the star systems within this volume, others are fainter stars. The Sun seen from 50 light years would be a magnitude 5.8 star, a faint star but not a red dwarf. Credits: atlasoftheuniverse.com.

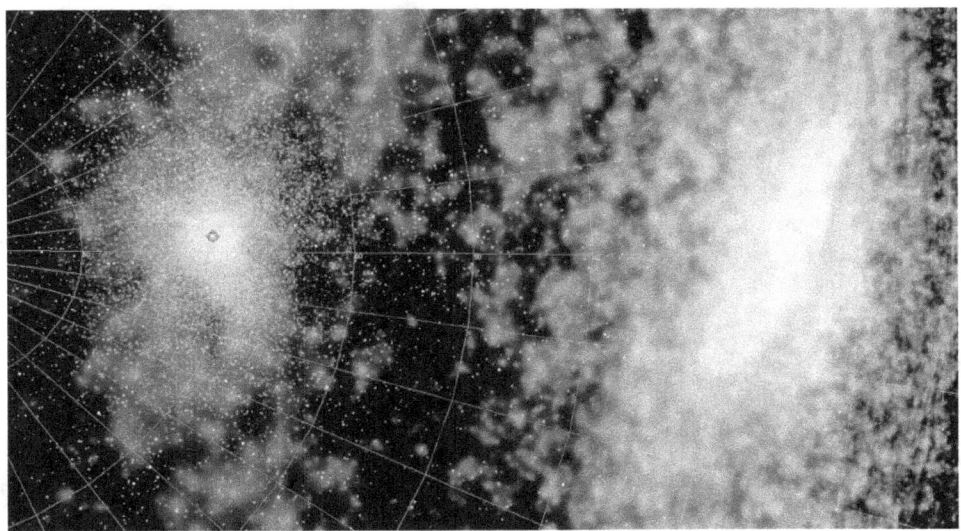

Figure 30 *First Cosmic Sky star density- Auxiliary view of Milky Way.*
GAIA Image- Orion spur, Milky Way; Star density in Solar neighbourhood; Submitted via GAIA Sky software, version 1.0.4, 2016. This effect is predominantly seen because of GAIA has graphed stars more clearly in the near neighbourhood. Credits: GAIA Sky software.

Figure 31 *Inside view of First Cosmic Sky.*
GAIA Image- Star density in Solar neighbourhood; Submitted via GAIA Sky software, version 1.0.4, 2016. Zoomed view of the upper image. Star density is seen to increase and concentrate, headed towards Sun. Credits: GAIA Sky software.

Figure 32 *View of First Cosmic Sky from within Solar system.*

Moon is not visible in frame, due to small size and dark side. Sirius (next to earth), a prominent star, nearest to earth in Canis Majoris constellation. Line diagrams show configuration of Astronomers consolidated constellations. Background surrounded with high star density- Submitted via GAIA Sky software, version 1.0.4, 2016. Credits: GAIA Sky software.

Figure 33 *Monarch Butterflies shrouding trees.*
Credits: WWF-Canada, awsassets.wwf.ca.

For your rebel some nerve, you're again enticed for a contest, to chance assess your bold vows; Allah knowingly has measured your true worth!

Allah the Almighty Lord is not the victim of failure. In His domain He only issues a command founded on the attributes of His Mightiness, Knowledge and immensity in Wisdom: *Quran 16:40 "For to anything which We (Allah) have willed, We but say the word, "Be", and it becomes."* Then things result in the making. A similitude is that of the many grains or fruits, which sprout in them are finer yet others still excellent. And He chooses to perfect from it as He wishes. Then, He assumes to His throne having brought about His intended design. A range of challenges are given in the Quran to deniers of God Almighty to prove their case. It calls them to produce a small chapter or even a single sign as narrated in it to prove the claim of Muhammad self-producing it. This challenge is not yet met. But apart from language for non-Arabic recipients the challenge is to deal with its narrative substantive to pick any discrepancy in it: *Quran 4:82 "Then do they not reflect upon the Quran? If it had been from (any) other than Allah, they would have found within it much contradiction."*

The range of challenges are linguistic excellence, sciences from a wide range of fields, a staggering straight forward challenge to create life or evade death.[195] As though Muhammad knew at the turn of our age researchers would be experimenting with Drosophila he wrote a challenge to create a fruit fly: *Quran 22:73 "O people, here is an example, so listen to it. Indeed, those you consider (entities) besides Allah will never create (as much as) a fruit fly, even if they <u>gathered together</u> for that purpose. And if the fly should steal away from them a (tiny) thing, they could not retrieve it from it. Weak are the seekers (followers) and the sought (Atheistic naturalists)."*

Next here is the challenge given to evade punishments sent by God. Western policy to destabilise middle east for oil such as in Iraq and now for upended strategic interests in Syria thru lies spread for anarchy by arming rebels with weaponry is akin to an extension of colonization which has ravaged the Arab world into a ghastly killing spree. This has led millions into destitution to beg refuge with neighbouring countries as the world saw it; it did mount a crisis at the European borders. Not knowing the law of karma that God will eventually equal justices: *Quran 2:134 "Such is a nation which has passed on. It had (the consequence of) what it earned, and you will have for what you have earned. And you will not be asked about what formers did"*, American people must at least now desist from age old destructive presidential legacies in foreign policy. The Arabic for karma is كسب (kasaba) which means "your earnings" for individuals and societies alike. This concept appears tens of times in the Quran reminding men and women to heed for warnings from God.

Further a specific challenge is given regarding astronomy. To cite a design flaw in the blanket surrounding earth. In the following sections the challenges from Allah are enumerated which are relevant to Atheistic naturalists in general and then some specific cases to astronomers and weather scientists.

---

[195] Quran 67:2 "(Allah) Who created death and life to test you (as to) which of you is best in deed - and He is the Exalted in Might, the Forgiving."

8.1 LIFE AND DEATH

> Quran 46:33 Do they not see that Allah, who created the skies
> (earthly) and the earth and did not become weary in their
> creation, is able to give life to the dead? Yes. Indeed, He is over
> all things competent.

When in today's science we are talking about de-extinction of lost species
how dogmatic we stand with our doubts concerning God's claim to resurrect us?
Things that are determined come about gradually. Allah is the most Patient and
therefore advices that patience is a virtue for His servants. In a verse where He
explains that a day for Him is that of a thousand years of our count; He thus
reminds us to have patience. But sceptics are in a haste to call on Allah's final
judgement i.e. to see truth right away: meaning immediately. They fall short to
realise that the design is drafted upon nations of human beings rather than a pool
of ingrates. Until the final time, many will Allah take by causing their lives to
reach death individually or in ways he has ordained. As the creation of earth and
its skies did not result in a failure; so, what prevents humanity from thinking that
Allah won't resurrect us back in a way that He created the planet and us for the
first time? It is a binding promise upon Himself which He will undertake and
achieve it far from succumbing to failure or frustration.

> Quran 8:50 If you could see, when the angels take the souls of
> the Unbelievers (at death), (How) they smite their faces and
> their backs, (saying): "Taste the penalty of the blazing Fire."

Among the many pretences of human mind is to get final answers from a dying
person; wondersome if they could quickly narrate stories of the netherworld
before dying. In such folly we risk sending our dear ones to guillotines, pre-
emptive of faith expecting answers. As it is... in reality spectators are told
nothing. Given to prejudices we say that none from the dead has thus far said
anything of where they were taken. And so, disbelievers dash the doomsday; their
spirit of enquiry strangely muting itself in this case. In fact, when the Angels for
death arrive and become permeable before the dying ones Lo! they can't speak a
word in our dimension. As the dimension of the living and its fleeting reality is a
done deal. The dying become dumbstruck and Angels will have begun taking
accounts and we would be helplessly oblivious to notice the catastrophe unfolding
before the dying as they step into death: *Quran 50:17-19 "When the two delegates*

make contact, seated on the right and on the left. Not a word is uttered by him (Man) but with him is a prepped bystander (to record). And comes intemperance of death in truth; (given) that is what you were trying to avoid." And on the other hand, those who have come to know the truth and humbled before God at death will be meted with good tokens and salutations by the Angels, in peace will they depart from life and in peace will be their resurrection: *Quran 16:32 "(Namely) those whose lives the angels take in a state of purity, saying (to them), 'Peace be on you; enter the Garden, because of (the good) which you did (in the world)'."*

> Quran 56:83-84 Then why do you not (intervene) when (the life of the dying man) reaches the throat. And you the while (sit) looking on?

> 85. But We are <u>nearer to him</u> than you, but you see not.

> 86-87. Then why do you not, if you are exempt from (future) account. Call back the soul, if you are true (in the claim of independence)?

Verse 56:85 tells that Allah is nearer to a dying person than we but we cannot see those Angels working with the dying person. Thus, God operates through His army of Angels in governing many subtle affairs on earth.

Modern day physics and biology agree with everlasting life. Biologists hope that one-day stem cell research will have advanced to de-age living tissue although reversing age of an organism is another ball game. Physicists mused by such miracles of biology explode black holes will anyway eventually make a morsel of us despite the leeway from biology. If the assertion of Creator's nonexistence was valid; then nature must have resorted to ways of immortality. By the powerful pilot of natural selection evolution must have at least appended longer lifespans which is a far easier proposition: otherwise, to lead a sea creature to mow the grass and then to return him back to waters to swim breathlessly long for making out of it a Whale after several hundreds of millions of births, preposterous! Has natural selection not realised that more than flying in the air in the limited altitudes of atmosphere; creatures wished to live for aye! Perhaps it could now begin to wonder in this direction after having worked so hard to make us live short. This argument is necessarily valid because all determinisms that went in to create life from non-existences would have won immortal life on the premise to live and not die. If this not be the case, then say, why would scientist striving to recreate origins of life end up creating death? Having imparted precious life unless death results by accident. Therefore, if life has by itself sought a way without a Creator it must have necessarily succeeded in immortality and this trait then must become a foundation trait among all species as evolution is marked to propagate lasting and populating traits. Many are dead when least expected. Young and handsome at home can pass away. Nothing close of an intended desire or imminence of a murderous plot neither any affliction by a suspended calamity but a silent mysterious death still occurs. Although we may attribute some biological reason for death after autopsy which is true. But in this age of medicine after one stand cured by the weirdest monstrosity known as cancer, and then to die of something like pneumonia at a young age in the wake of measures to increase the average lifespan sucks!

> Quran 56:58-60 Do you then see? The (human Seed) that you throw out. Is it you who create it, or are We the Creators? We have decreed Death to be your common lot, and We are not to be frustrated.

The indicator in the Quran that the DNA (human seed) is loaded with codes to die is given to frustrate us. As it points owing to its intricate code that we would not be able to sort it- to live an immortal life is clearly a teaser. These are the kind of breakthroughs which are expected from stalwarts of science who want to press their opinions sought by folly to deconstruct faith. If they can achieve freedom from death, there you go, they would demonstrate true claims to science as liberation from the darkness of dying. Why must a man after attaining great heights in knowledge and having eaten organic all his life die? Why mustn't he live for the coming greater times to become the immortal legend like the presumptuous jelly fish that shares a common ancestry, from where vertebrates are presumed in scientific literature to have branched out; splitting very early in the acclaimed evolution path. Demonstrating this some trees are given to live for thousands of years: Bristlecone pine lives for 5000 years. Some sea creatures such as the ocean quahog live for 500 years. Then another early in evolutionary chain, ocean sponges such as calcifies demosponges are estimated to live for thousands of years. Why only some creatures? Whereas numerous species lag from acquiring- the most wanted gene. From what the Quran comes to give guidance in this area of sciences I have found Darwin's observations in the limited context of variation of subspecies commendable; however, this segment of argument is not by natural selection as misunderstood instead per the Quran it is for natural ecosystems. Darwin's exaggerated extrapolations regarding branching of all of the various kinds and species from common ancestors amounts to delusions. Natural selection has clearly deselected the golden mantra of long-living, if not immortality. Yet allegedly perfecting all other traits to carve the most capable machine- the human; awarding for most a life filled with hardships as told in scriptures. A sarcasm shown by naturalists as injustice on God's part for our many sufferings in this life: *Quran 84:6 "O' mankind, indeed you are labouring toward your Lord with (great) exertion and will meet Him"* after Adam and Eve had before dwelled in an immortal felicity.[196]

Bringing forward the most important traits for survival was long thought and swanked as the essence underpinning the philosophy of survival of the fittest and natural selection. Further, why must experts in fields of science and research ever give-up from fighting to win over death? It is shaming to succumb to death after orchestrating many massive claims for freedom of life. God tellingly has reminded us that we aren't freed from His grapples of death: *Quran 90:4-5 "Verily We (Allah) have created man into toil and struggle. Thinks he, that none has power over him?"* In this age of gene decoding it is incumbent upon claimers to free themselves from the clutches of the death-bringing God and employ smart ways of gene-editing to do away with death instead of the lazy proposition of deep-field observations in the gorges sky in the depths of the universe and concluding that we're still barely able to see. Then to recline back "no, we haven't spotted the Lord", perhaps Math equations with integral of Him aren't faring well. They assert given the Creator did result in a creation; but how or who caused His first being still does not fit in the Math. This is an impulse at which they attempt to dissolve the Uncreated Supremo. He, God our Lord Allah is anxious and is yet very patient. He Almighty is ardently teaching us that the whimsical sequence of infinite ingress in not a proper Math. Therefore, in my humble opinion the sceptics are better-off to learn afresh in good schools.

---

[196] Quran 20:117 So We (Allah) said, O' Adam, indeed this is an enemy to you and to your partner. Then let him not remove you from Paradise so you would suffer.

8.2 TO EVADE EARTHLY PUNISHMENTS

> Quran 23:17 And We (Allah) have created above you seven **tracks** and never have We been of (Our) creation unaware.

In this verse it is clearly spelled that seven earthly skies are a means to extend the reach of Allah upon earth. Each earthly sky can cherish and protect life as much as it is equally able to chastise and destroy life. The Arabic word طرائق (tauraayiqh) translated pathways or tracks is google translated as **modalities.** Each of these skies are a means to reach earthly residents. Atmosphere besides sustaining and providing for life brings in severe weather; Ozone besides life protecting, it's pungent and obnoxiousness can kill and when thinned down can pass in harmful band of UV radiation; Ionosphere besides aiding in communications has complex roles, it partakes in atmospheric electrical activity and is reflexive to fissures across earth; research is on. Plasmasphere and Radiation belts besides guarding the earth's surface can channel high energy particles and solar storms upon earth. Magnetosphere if dysfunctional may lead to the blanket of atmosphere surrounding earth fade away due to solar activity. As its significance stands unparalleled. And the satellite sky could attract its own swarm of dangers from the depths of space with its looming meteors and asteroids as it did in its distant past. The verse ending **"and never have We been of (Our) creation unaware"** subtly alludes to the earthly skies being also a means to keep His creation (men and jinn) in check.

> Quran 13:31 ...and disbelievers do not cease but be struck with disaster for how they have acted. Or that calamities must descend **near their home-** until there comes the final promise of Allah. Indeed, Allah does not undo His promise.

If we only pay attention to natural calamities across the world descending at our door steps it is exactly what the Quran is informing. It was not for a human in seventh century to narrate his comprehensions of wide scale natural calamities unfolding and scripting of stark details gathering responses deep from human psychology. There is not a cyclone or a hurricane whose aftermath appears as spooky as a war-zone; yet this does not desist us from the meme of a brute. All such events are a sign from Allah as smaller punishments for evidential sin before the inevitable judgement of God must occur. Alas! if we could repent and supplicate for God's guidance in which case His mercy is guaranteed—it can't be simpler said than this. The meme of being superior is so strong that it rubbishes holy restrictions convened by faith upon humanity; like the state Riyadh in its tyrannical deliberations upon its neighbours bereft of human values. The law of karma, spiritual forces at play for justifications, will eventually prevail. Given a clear surge in empirical data of many destructions by natural disasters across the world; these won't still suffice science con men and pseudo religious to comprehend reality telling by God. In this brute culture of superiority and race swelled with ill intents of amassing resources a geopolitical mischief culminates and mayhem surges within societies leading to false policy infiltration of nations. Must this then surprise us that God's turning of knobs to release back to back hurricanes are not given to punish us? Harvey and Irma are like the beads let loose from a string to reciprocate feelings of deprivation at least in the context of Syria so that nations may heed, and others may heel. But seemingly nothing of this is coming to our good realisations. Introspecting into our sinful past and current oppressive world affairs it is possible to foresee our future predicaments with weather and earth in that it may perhaps help us to pursue ways to avoid its fury. A great nation can do things to move God by great measures for overall good. It is apparent that we need to blame our moral actions more than our industrial pursuits for several calamities. Morals also straighten industrial practice for our being content with little, making industries leaner, greener and efficient. Seemingly not learning that Allah has helped mankind progress all along from the dangers of the earth and sea; being ungrateful for His many blessings such as the awarded intellect, we heedlessly churn towards our destruction. So, when we commit atrocities (Arabic: fasad) on earth, unmindful of accountability to a higher power not even acknowledging Allah's existence, His displeasure surrounds us by His punishing countenance turning the angry face of mother nature against the trespassers. It is a pointer I'm going to lay evidence for.

> Quran 16:45-47 Then, do those who deploy evil policies feel secure that Allah will not cause the earth to swallow them or that the punishment will not come upon them from where they do not perceive?
>
> 46. Or that He would not seize them during their (usual) activity, and they could not cause failure?

47. Or that He would not seize them into gradual wastage? But indeed, your Lord is Kind and Merciful.

It is a well-known fact that Almighty Lord Allah in Islam is the Lord of Compassion, Mercy and Love. He is without hesitation the Lord of Justice which require enforcing law and retribution to right the wrongs. For the mere fact that His creation owes Him acknowledgement and servitude. As He says, He would just do away with those who deny knowing Him. Per the Quran any calamity to befall men has split possibilities. Firstly, one that comes to test and assesses mankind their beliefs and unwavering faith despite the ensuing sufferings day in and day out to differentiate friends to God like Prophet Abraham from commoners in faith and then others unworthy of faith. Secondly, that it is unleashed as a chastisement in this very world for our sins. The latter having to do with Allah's displeasure of our weak or unworthy faith and for our mistreatment of fellow Human beings. Apparently, the many droughts, cyclones, hurricanes, tropical storms, devastating hail, floods, earthquakes and fires across the world are telling of these Quran verses. The severe storms that we are getting to witness aren't just the tail-end of a deformed climate model. If it was so perhaps high energy earthquakes hadn't followed suit in similar exhibition as it stands clearly explainable that these two sources of energy dissipation are miles unallied. The premise that 1°C rise in global temperature is solely responsible for violent weather is questionable. Storms and other inflictions are from Allah. Allah would still send it our way despite temperatures plummeting to historic global normal. The surge in greenhouse gases wasn't necessary for times when Noah and his companions were carried in the Ark as it rained and stormed for many days and drowned them utterly, flooding the landscape and thoroughly encompassing the population of his people who domiciled there. Per the Quran the interpretation of Noah's flood is that of a localised occurrence as already discussed in section 4.2. The world for Noah's people was confined to every bit of where they had lived and travelled for their necessities in those prehistoric times. Then Human population was all that belonged to Noah's civilisation.[197] And

---

[197] Quran 36:41-43 And a sign for them is that We (Allah) carried their forefathers in a laden ship. And We created for them from the likes of it that which they ride. And if We should will, We could drown them; then no one responding to a cry would be there for them, nor would they be saved.

neither did the weather during Joseph's tenure in Egypt need Global warming or other anomalous conditions for the drought had lasted for seven years. As go other examples in history.

It is understood that rise in global temperature helps surge the troposphere with increased evaporation of water in a single summer.[198] Which means more rain fall is to be expected, but can it be fully justifiable to allude this towards violent weather making landfalls on human colonies? Winds are responsible players in transporting clouds as they form. I wonder if these winds are playing a discriminatory role? Winds amass and move heaps of storms worldwide. As frightening as it is they seem to have driven massive Hurricanes very deep in-land in regions of American plates.[199] Over the course of 20th century from 1900-2015 all 50 states in the USA have receipted a worst category of Hurricane. Mother nature speaks as if she is very calculative!

As argued therefore Allah may choose various ways to warn us even punish or perhaps give us some more time to reflect; but He has explained that these are the ways that He chooses to approach mankind when He seeks retribution.

> Quran 21:11-14 How many were the populations We destroyed because of their iniquities, setting up in their place other peoples? Yet, when they felt Our Punishment (coming), behold, they (tried to) flee from it.
>
> 13. Flee not but return to the good things of this life which were given you, and to your homes in order that ye may be called to account.
>
> 14. They said: "Ah! woe to us! We were indeed wrong-doers!"

Today Global Warming is a valid issue and we do need to badly worry about it. As it is clear from the **Figure 34** temperature anomalies are true; the pursuit of materialism has pushed industry for a rapid race in expansion. Led by stock-

---

[198] Per Climate Council of Australia in a research on Super-charged storms in Australia: The Influence of Climate Change by Professor Will Steffen and Dr David Alexander reference a quote: That water vapor content is estimated to see 7% increase per 1°C warming; Trenberth 2011.

[199] The data on weather.com/storms/hurricane, records worst category Hurricanes to have hit every state in the USA.

piling and boasting; mankind's chase for economic well-being and grandeur has brought us to the fore on containment of greenhouse gases. It is also clear that we need to adapt fast to be industrially clean and lean. The graph **Figure 34** here shows temperature anomalies which is an indicator of rapid industrialisation since mid-1900s. It is not surprising that the weather has taken striking resemblance in that to say frequency and severity of charged weather has followed suit. But what remains incomprehensible is even earthquakes show similar patterns. Has global warming influenced earth that deeply? The earthquake data can be measured for destruction based on unique factors specific to geographical locations and population domiciled in those regions which are main factors that scale up the measure of destruction, pointing to the type and age of infrastructure and population densities. A magnitude of 7 and above on the Richter scale is taken to discussion. The first and the second half of 20th century have recorded 90 and 106 counts of earthquakes respectively. The 21st century in just 16 years thus far has experienced 252 counts, data finalised up to December 2016.

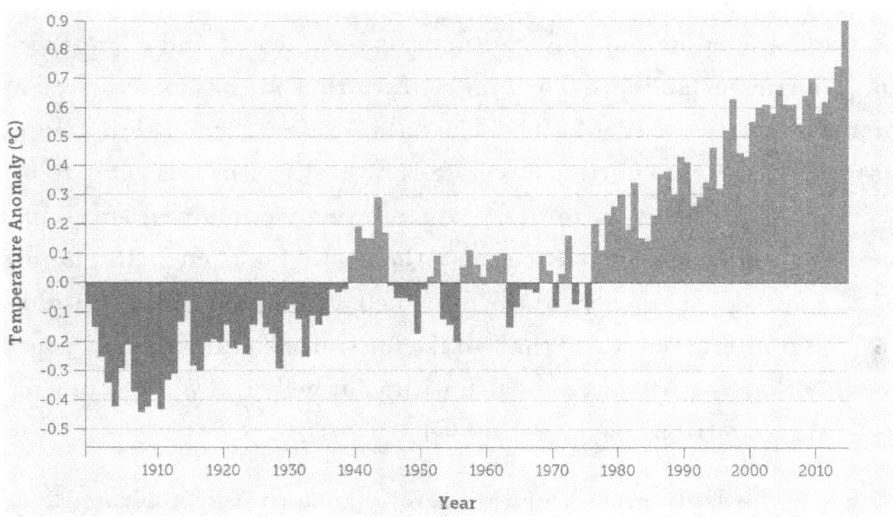

Figure 34. Annual global temperature anomalies.
Relative to global annual average temperature 1901-2000; Credit: NOAA 2015.

On the other measure the high energy earthquakes of seismic magnitude 8 plus stand at 51 counts in the 20th century. But, surprising as it is for projection, being in the beginning of 21st century, thus far more than 20 counts of high energy crust crumps have already been recorded.[200] Naturalists to dispose of inferences on God's hand in it also withhold data on cluster of earthquakes from reporting. With local clusters of aftershocks making up to 1089 counts being removed from a total of 2200 counts, this amounts to nearly 50% of data being dismissed from plotting. Such indicators removed will clearly flat out the peaks of the curve in the graph. For example, on 26 October 2016 a 6.1 earthquake struck followed by a 6.6 on 30 October 2016; both struck significant damages in the regions of Italy and were aftershocks from an earlier 6.2 earthquake in August 2016. Three strikes bringing destruction were reckoned as one. Which shows they are subjected to deliberate census cut-downs. Much of the numbers removed due to such de-clustering that have accrued over the last couple of decades is indicative of crafting to suppress the reality of increased responses from the earth.

With Storms it is very difficult to classify and report as there are hurricanes, typhoons, heavy hail storms, cloudbursts and flash floods, cyclones, desert dust storms etc. For sanity purposes I have analysed data from the history of worst hurricanes of the tropical Atlantic storms. As this narrows our research area to the Atlantic basin with landfalls in North and South American plates; this approach thus makes it clear to realise the impact surprising audience worldwide. The first and second half of 20th century has recorded 11 and 53 categorised as the worst hurricane landfalls. The 21st century thus far has experienced at least 50 worst such hurricane landfalls.[201] The data shows clear four hundred percent increase in catastrophic hurricane landfalls in 21st century as compared to the 20th century. This helps us to grasp a good understanding of surge in severe weather. If Humans do not treat others the same, for what is the natural law spelled by many good teachers of the past such as Moses, Jesus and Muhammad for the greater influential world that—Like for your neighbour, what you like for yourself and that you will have no faith with your belly full while your neighbour starves—Allah's retribution is guaranteed.

Observing the birds fly in and migrate across lands to fulfil their needs, plainly reveals vital wisdom that we aren't to prevent them or the refugees drifting for

---

[200] The statistical data is sourced from Wikipedia and from United States Geological Survey. 21st century data as compared to 20th century engagement of earthquakes. [201] Data sourced from hurricaneville.com/ historic.html; weather.com/ storms/ hurricanes and Wikipedia.

life. Their rights to traverse the landscapes for living are credible. Natural mechanisms are conscionable unless violated by human greed such as nationalism controlled by repressive immigration policies. It is wise to know that God is the Cherisher of the worlds. Seeking Him, pleading His forgiveness & honestly pursuing justice will save us billions of dollars spent in revamping infrastructure post dreadful destructions and the money expended to detain refugees in impoverished lands. As argued before, therefore Allah may choose various ways to remind us or even punish us or perhaps give us some more time to relish but He has explained that these are the ways of Him to rebuke mankind when He seeks to avenge on behalf of the oppressed.

8.3 MISUNDERSTANDING SCIENCE

> Quran 67:3-4 (Allah is He) Who created seven-skies (earthly)
> overlapped. Do you see in the creation of the Most Merciful any
> inconsistency? So, return (your) vision (to the sky); do you see
> any flaw? Then return (your) vision twice again. (Your) vision
> will return to you humbled while it is fatigued.

Planet earth's blanket sky is convincing and coherent. Also, the Cosmos
down to earth is powerful and bold. It is exactly as it is described in the Quran
that our vision returns fatigued upon details of observations in the blanket
surrounding earth and that of the deep celestial space. Lately, in science circles
the use of scientific language concerning earthly skies is rather belittling of the
powerful design surrounding us. It has been inferred in language that
Magnetosphere yields at the cusps. A phenomenon known as Aurora borealis in
the north and Aurora australis in the south is alluded to leaking radiation (see
Figure 35 and **Figure 36**). The magnetic poles of the northern and southern
hemispheres also called cusps are taken as a couple of rifts in the design of earthly
Magnetospheric sky. It is further construed after the Quebec Blackout caused by
a solar storm that Magnetosphere is penetrable![202] By this view it is possible that
the claim made in the Quran on consistency in Allah's creation lacks
corroboration. At least from the thoughtless documentarians who often resonate
beams of harsh language to split a crack open in the profound design of earth's
blanket sky. The thoughtlessness in their ability to language and reason is
miserable. Seemingly pathways for sustenance and retributions alike are spoken
of as rifts in the sky. Forest fires may be spoken of as destruction but to its
immediate coming burnt ashes provide for another epistle growth. Earthquakes
is again another means of reachability to punish us. Here, critics may cite the
Quran for a stable, comfortable and bed-like earth indicative of many verses and
then say quaking is therefore a flaw in Allah's design. Similar things may be said
of devastating cyclones after the Quran has spoken of the earthly sky as a canopy.
If a solar storm hits the earth it does not mean that Magnetosphere is a bad design
or has a flaw. Aurora's are not a rift in the design as seen and talked, but it is a

---

[202] A NASA blog in an article titled: The Day the Sun Brought Darkness; Excerpts
from Dr.Sten Odenwald, NASA Astronomer, dated March 14, 2009 and last updated,
July 31, 2015 by Holly Zell.

phenomenon unfolding in the transit areas of a seamless marvel allowing solar essentials to integrate hosting the earth's blanket skies. The wisdom of Allah supersedes the narrow thought line of naturalists. To critique Allah's artistry in the design of skies it must not be mere verbal deliverances. If there were a rift it would therefore call our ability to propose an engineered alternative. A design proposal that identifies rifts in the present and tenders a mitigation which would suffice a better design. But then there isn't simply another alternative; Magnetosphere is just too cool and bold.

In fact, the design plan of Allah is retentive of His ability to reach inhabitants on earth. Abnormal transients experienced from space weather is the reserved ability of Allah to demonstrate His capacity for retribution. One among the laws of Allah is His wanting to punish us in this very world upon citations of disbelief and debauchery. It is also a channel of justification to show Allah's evidence to believers of accounting disbelievers in this very world. But on other hand, flaws are inhibitors to normal operations which when loaded with transient forces they shred the embodiments of design into rumbles. Magnetosphere does not yield to any discontinuities which is evidence enough for its design perfection. There is just one way out for us, submission before the Creator—which is exceedingly clear.

Another talked subject is of the South Atlantic Anomaly (SAA) near to the Antarctic continent in the southern hemisphere. It is an area where the earth's Van Allen radiation belt listed in this book as sixth sky comes closest to the earth's surface. Dipping down to an altitude of 200 kms. It is simply one among the many evidential findings of the Quran talking of the earthly sky design as an overlapped architecture. Explaining the design for sourcing of plasma in its early phase of construction from the sweltering of oceans which were rife due to global proportions of volcanic activity. Therefore, it is not an abnormality but clearly an essential element of finalising the overlapped design.

Yet another earth's sky-bound phenomenon misclassified are the cloudbursts. Cloudbursts are known to precipitate large sums of water suddenly and dumping them creating flood conditions. On 15 June 2013, such an event gripped Uttarakhand in India were 5,700 people were presumed dead including 934 local residents (Wikipedia). Several cloudburst events are only recently reported due to increased awareness concerning abnormal weather. In fact it is likely that it is a common phenomenon and less reported. However, it has a historic track record of inflicting human populations. The Quran classifies this event as falling of a piece of sky.

> Quran 52:44-45 And if they were to see a fragment from the sky falling, they would say, "(It is merely) clouds heaped up." So, leave them until they meet their Day in which they will faint.

Each square inch column of air above earth's surface weighs 14.7 lbs. The total mass of earth-bound atmosphere is about 5.6 quadrillion tons. Lumping of atmospheric mass which may be caused by the localised freezing of atmosphere is the way Allah brings down a topographical sky bearing clouds upon a region dumping the cloud mass. Clouds then rapidly precipitate around dust nuclei near surface of biosphere causing sudden release of water in a region.

> Quran 34:9 Don't they see what is before them and behind them, of the sky and the earth? If We wished, We could cause the earth to swallow them up, or cause a piece of the **sky to fall upon them**. Verily in this is a Sign for every devotee that turns to Allah (in repentance).[203]

Critics not realising science behind sky falling may say such a thing is obviously impossible as the earth's atmosphere is simply made of gases. Which appears like horse-lensing ideas of science. Are they then to wait for the inevitable to believe if Almighty's narrative were undeniably true? Atheistic naturalists who secede at Almighty Lord's threats of hell say: How could the Lord of worlds be bankrupt of -convincing statements-, that He has to scare us by hell for belief—isn't this obnoxious of an Almighty? The irony is, these are the very people who help educate pupils that one day the sun will scorch the earth's surface blistering its inhabitants! Nevertheless, they are unable to see punishments of Allah or even His telling of the most imminent probabilities that lurk in near celestial horizons to map our fate, such that they can desist from misbeliefs.

The famous god of love Venus is ample of a hell. Lurking in the evening and morning sky showing the face of love is a chattel of deception. Such is the waiver, if Allah does not bestow favour we will be left to desolateness in the dying earth;

---

[203] Also, Quran 22:65 ...and He restrains the sky from falling upon the earth, unless by His permission...

not to have a God that rescues is indeed harsh! The delusion is, naturalists winch inferences of Almighty Lord as being a cruel entity but then they freely talk about the merciless natural phenomenon and an uncaring Universe that doesn't give a damn to share some love... Heavens! Atheist naturalists for their good could realise that God was more caring of us than our mothers! It just takes a single cry of supplication to Allah lest He has openly owed to punish.

Our understanding of science is yet to classify many natural phenomena in their correct lenses. Earthquake is clearly one such phenomenon whose root cause is too complex to pinpoint. Though at one-point scientific understanding was very suggestive of earthquakes occurring due to plate tectonics. But findings of earthquakes within tectonic plates has rendered our understanding of it as insufficient.

Regarding -planetary orbits- once imagined and acclaimed by scientists, their mathematics to revolve in near round trajectories around the sun today are thwarted by new science revelations; which mock at our miscomprehension of this feature and the misguided notion that this is a necessary primary behaviour of planets. Today scientists note that earth like orbits are extremely uncommon. Our understandings of how a planet must behave is therefore challenged. Sedna in our solar system along with several other dwarfs and comets orbit the sun in highly eccentric ellipses (see **Figure 25** and **Figure 26**). In fact, astonishingly modern science reveals near circular orbits aren't very common in universe at all. Many planets orbit their stars in oblong orbits such as HD80606-B a gas-giant 200 ly away in Ursa-major constellation. It comes close to its star 3 million miles and wanders away for up to 78 million miles traversing in extremely eccentric ellipse every 111 days. Our earth in near round orbit drifts about only 4 million miles. Then, what possible explanations drive these extreme eccentric orbits? Apart from speculations of gravity perturbations caused by a passer-by object, such as a star. However, speculations of the perturbed orbits for the likes of Sedna in our solar system due to its long orbital time of 11,000 years is quickly churned away by the latter mentioned gas giant in Ursa-major with its short orbital periods around its star. Traversing three times for every earthly year it is seen refusing to normalise from its perturbed orbit. Telling that stories of perturbations are only an illusion![204] What is the force otherwise keeping Jupiter-sized planetary objects such as HD80606-B in Ursa-major in unrelenting ellipses? Can gorging of anomalous orbits in celestial space be a demonstration of never-ending mysteries. Also, are the kozai satellites perturbed by the moon. Scientists have already put this concept to functional use by reducing this to a three-body problem via idealising by Kozai mechanism satellites. These are made to linger over one part of earth for a long time and then they fly-by closely around the other side of the earth perturbed by the Moon. These GPS satellites are held in elliptical orbits to maintain good coverage of earth round the clock.[205] This does not explain the God of gaps hypothesis in case you're mistaken by this piece of narrative alluding to anomalous behaviours. Even if these mysteries along with several unknowns such as the nature of gravitation itself were fully explained by our endeavoured scientific inquiries yet God's plausibility is not compromised. God's existence is not detrimental to how much truth we comprehend. Our understanding and growth could be endless, on the other side **truth has no beginning** in the larger world; it has always existed. God, -the truth- is the first and the lasting to whom all matters are referred: Quran 57:3 *"He (Allah) is the First and the Lasting, the Evident and the Immanent: and He has full knowledge of all things."* Unfortunately, westerner theoreticians are prone to concoct concepts. They assign funky words in an effort to breathe life into it such as "the God of gaps" argument. But in eastern cultures at least in my upbringing we were never taught to see God in this boundary. Instead we knew that we can always fault, and God is absolute and thus our faults and limits do not bring God's existence into questioning. As it is popularly said of Jesus Christ: "seek ye the truth and the truth shall set you free" is the norm in God's religion. The Quran is entirely based on encouraging human reflections, prophesying that such humility to earn truth is the sign of the servants of most Gracious Lord: *Quran 5:83 "And when they hear what has been revealed to the Prophet, you see their eyes overflowing with tears because of what they have recognized of the **truth**. They say, "Our Lord, we have believed, so register us among the witnesses."*

---

[204] Watch on YouTube, Traveling to Other Galaxies - The Search for Earth like Alien Planets - Space Documentary; published 23 Feb 2017. Video time: 27-30 minutes.
[205] Definition of Elliptical Orbits by Zachary G. Brown; Updated April 24, 2017; sciencing.com. Also see Kozai mechanism on Wikipedia.

Regarding -Heaven and Hell- naturalists aren't the first ones to confront these topics. As eager as ever they attest a card of science to assert triumph. The Quran has captured lengthy arguments of Prophet Muhammad's tribesmen debating afterlife with him centuries ago. When Allah said *Quran 6:33 "We (Allah) clearly are aware the grief their words do cause you: It is not you they reject: it is the verses of Allah, which the wicked contend."* In fact, the Quran informs this debate first stemmed in Noah's time and since then every nation except a few has repeated these standpoints. Therefore, today's cries aren't sought of science! They are but sought passionately to exercise liberty from Allah's judgement. Having not worked on evidences, how can this be wise of the bright minds to argue concerning eternal life? Satan had this very impulse, to do his own and not Allah's. He rebelled against Allah, citing Allah's favours on Adam, in this pretence he stung the charge of misguidance upon God as he turned away declaring to lead many of us into rebellion. Thus, it is affected by Satan: *Quran 15:39 "(Satan) said: "O my Lord! because You have put me in the wrong, I will make (wrong) fair-seeming to them on the earth, and I will put them all in the wrong."* Except the sincere ones, as properly clarified by God in the following verse: *'Those who, would have set a guard against you and would humble to Me would therefore I be their guide'*.

The objective of science is wholly misrepresented by dishonest efforts to tell that it contradicts the divine doctrine -speaking for the Quran- the final message from Allah. Science is a way of knowing and validating the known. And religion is a way of living. In fact, religion and I speak for Islam- the way defining human limits given by God. It hosts science and considers it a potent tool to benefit mankind. If naturalists cannot come to terms with this definition here then you must realise that half-baked science of the ancient Greeks had already made arguments what you now come to think as fetched by modern science and as we already know Greek heresy was replaced by teachings of Jesus Christ the son of Mary. Similarly, the Quran's narrative if not rebutted by naturalist community today; has an equal effect over modern heresy. Richard Dawkins must realise that generational upgrades in science is irrelevant to those who have died. It is only a lifetime that allows an individual to make a sure choice of his own regarding belief in God. Allah ever since has noted vagrancy in discourses on part of the naturalists saying: *Quran 52:12-14 "They (disbelievers) are in (empty) discourses amusing themselves. The Day they are called toward the fire of Hell with a thrust, (telling), 'This is the Fire which you used to deny'…"* then they will be asked if this still feels unreal to you or you still don't perceive it as you were denying it wholesomely? So, it will be said, reside here, your patience or impatience thence is the same.

The number crunching may be true to the last digit (millions of years) to sun's dying holds naturalists in good composure for now. So, they laughingly ask: when is God effecting this catastrophe? *Quran 51:12-14 "They ask, 'When is the Day of Recompense?' The Day they will be stricken to trail over the Fire."* It will be said: now home in your requests, Allah is not into idle sport! Quran *38:27-28 "And We did not create the skies and the earth and that between them aimlessly. That is the* **assumption of those who disbelieve**, ... *Or should we treat those who believe and do righteous deeds like corrupters in the land? Or should We treat those who fear Allah like the wicked?"*

Space endeavours in fact confirm the truth told in the Quran of a hell. Probes sending data from most planetary missions within solar system show more fierce resemblances of hell like situations where huge minaret like volcanic matter is ejected and where there is constantly charred atmosphere. How many more facts can be given in scripture to understand Lord's position. Almighty Lord keeps mastery of facilitating fiery things into reality and in another paradigm protecting believers as told:

> Quran 57:13 That Day Hypocrites men and women will say to those who Believed: "Wait for us! Let us borrow from your Light!" It will be said: "Turn ye back to your rear! then seek a Light (where ye can)!" So, a barrier will be setup betwixt them, with a gate thereunto. Within its realms will be Mercy and outside across its peripheries, will be (Wrath and) Punishment.

> 14. (They) will call out, "Were we not with you?" (Believers) will reply, "True! but ye led yourselves into temptation; ye waited and doubted (Allah's Promise); and your desires deceived you; until there issued the Command of Allah. And the Deceiver (Satan) deceived you in respect of Allah.

Giving a glimpse of the Day of Judgement, when earth will be gripped, and chaos will assume, the ripped atmosphere will be churned into zones where there will be a barrier of cool and shade into which believers will be admitted and pretenders (hypocrites) will be left alone. Those who had false suppositions of Allah and who were in pursuit of materials will be separated from the ranks of believers. Believers will shine forth a glow from their right hands by bioluminescence launching their torches and be in peace while the disbelievers will wander in turmoils. At length: *Quran 39:69 "…the Earth will shine with the Glory of its Lord"* when passes will be issued, believers will rejoice, and they will be securely escorted to gardens in Paradise planet. The disbelievers blinded, they will be flung into the pits of gloom hurling huge plumes and columns of smoke: *Quran 77:30-33 "(They be said) Proceed ye to a shadow (of ashes) in three columns, no shade of coolness, and no help against the Blaze. 'Indeed, it throws about sparks (as huge) Forts. As if there were (a string of) yellow camels (marching swiftly)'."* Drawing scenes from space probes, this Quranic description of what a hell contains is apt. Smoky columns of volcanic ashes will not provide any shade but be constantly throwing lava, like a line of yellow camels seen from afar. Should one wonder at its factualness? Or should one wonder as to why it is not frightening? This is the real picture before us, making it a matter to contemplate as we now know what is befalling us. Thus, none can make a claim of him being lost in this mysterious Universe without any purpose. Then for those of submission, felicity is in que, being united with friends and family in an afterlife. A loss of this and condemnation to hell is unquestionably a manifest loss: *Quran 39:16 "They will have canopies of fiery smoke above them and below them, canopies. By that **Allah threatens His servants**. O My (Allah's) servants, then fear."*

> Quran 17:58-59 And there is no city but that We will destroy it before the Day of Resurrection or punish it with a severe punishment that has ever been in the Register inscribed. And nothing has prevented Us from sending signs except that the former peoples have denied them...

Clearly Moses's performance of miracles today is a popular myth among sceptics so is Jesus's. The disbelievers surrounding Muhammad also requested for a miracle as a precursor to belief. He was told by Allah to fetch one if he could by himself.[206] Allah was not to show any such miracles of the past only to be further idolised in select circles and denied by many as uninspiring and fantastic

---

[206] Quran 6:35 And if their evasion is difficult for you, then if you could seek a tunnel into the earth or a stairway into the sky to bring them a sign, (then do so). But if Allah had willed, He would have united them upon guidance. So never be of the ignorant.

malarkeys propounded in religions. The plain miracle that was given was the Quran; to last for rest of the times until doomsday. It was the Quran that was pointed to as the unfailing argument of Allah for deniers to refute faith. But, naturalists little understand revelations! Granting they insist that Allah must be sending revelations for modern times. Allah could have, but if modern man could request intelligently. To their dismay due to advances in information reception, then whomsoever had antennas would accede to be honorific of a Prophet, silly as it sounds. Prophets are not iconic or be qualified to be Prophets simply by possessing stately arts nor do they make figures of glamour, but they are instruments unparalleled of fine mannerism, character and calibre among the populace. It was the time with Muhammad that Allah deemed fit to seal the office of Prophethood, so He did. And the believers in Paradise will converse 'and indeed, we remembered God as the Beneficent and Merciful': *Quran 52:27 "So Allah conferred favour upon us and protected us from the punishment of the Scorching Fire."*

In a subtle nature, our science-based enquiry hasn't still understood Allah's plan on earth to test its inhabitants of freewill for their belief and conduct. Freewill is not as Sam Harris [207] would like to bust it as a fallacy by invoking instrumentation to calibre -emergent thoughts- even prior to realisation within oneself. Equalling this to a great feat in human achievements is naive. You may not need this instrumentation to tell you 'those' at all. This happens knowingly with some awareness and self-reflection. When to face stage fear, prior to his turn the anticipant after a delay realises that his heart beats have increased. It is like the subconscious does not tell you things by the instant of its inception. And reflex mechanisms help evolve emergent outcomes and sedate comprehensions of these pulses rudimentarily helping the consciousness derive only necessary outcomes from all that is being processed. But knowledge of such unawareness's can be acquired upon inquiry and better perceiving of reality within and surrounding us, not inherently predetermined as Sam Harris argues. Any outcome of decision making which resides within deeper subconsciousness is in one's predisposition as the ability to reflex, but its evidential manifestation is nevertheless in one's control of consciousness. These get exercised with knowledge, practice, experience, and, above all patience which make essential

---

[207] Samuel Benjamin Harris is an American author, philosopher, and a neuroscientist.

tools in exercising freewill. Without these essentials, human equals to Animals wherein rudimentary laws govern determinisms. Therefore, Animals do not accord to adjudication giving pronouncements concerning justices among their kind. But humans are hosts to such attributes to exercise freewill.

Naturalists assert of living "highly ethical" not realising that they present as evidence many of their false desires: Quran 51:8 *"Indeed, you (denying) are in differed themes."* With a determined will to disengage from clear stage they rejoice within separatist circloids, thus they are misled. Naturalists spearheading flags of Atheism condescend God the Creator, by mere whims. Having not supported their claims to disprove God's existence at least by rationality. This therefore is evidence enough for their wrongs having become fair seeming to them. The element of just position in the least beckons their hardened nerves to be neutral concerning God where agnosticism is a better stage than Atheism.

Figure 35 *Aurora Borealis.*

ESA (European Space Agency) astronaut Samantha Cristoforetti took this aurora borealis Earth observation image from the cupola window of the International Space Station on Dec. 9, 2014. Image source: nasa.gov/gallery. Credits: European Space Agency (ESA)/NASA.

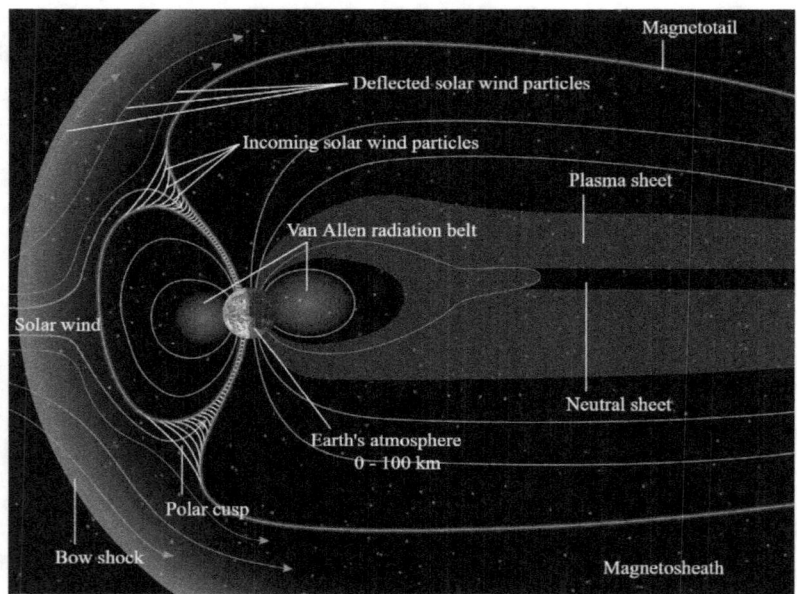

Figure 36 *Magnetosphere- Detailed illustration.*

Earth's sky boundaries: Magnetosheath and phenomenon down to earth; Day side is compressed, and Dark side stretch of skies is apparent in this figure, their (Magnetosheath & Plasma sheet's) effective stretch is up to 1000 earth radii (not shown); normally skies enclose Moon's orbiting at 60 earth radii. Image sourced: Wikimedia commons; Credits: NASA.

9 INDETERMINISM AND DOOMSDAY

Frustrating is lockdowns and chains. Thereafter-judgement it is equal even if you wail. For liberty is an asset to worship at will; again, payments are heavy, and stakes are high!

I think that you may be more sceptical at this point of reading this book about what is said of current quantum physics. If this emergent science will refute faith in God:

That would have us believe that there is no need for a supernatural intervention to have created the universe and us just as we are. The world feels warmer and more liveable if we believe we're part of something bigger than ourselves. Without it, it can feel cold, heartless and uncaring beyond our human relationships.[208]

After reading into this book readers in my known circle from the White community have openly expressed their desire, to see beyond; to look for clear evidence. Than just heartlessly critiquing the natural need felt for belief by the human consciousness. Now, I have to tell you at this point that it feels equally cold despite belief. As belief makes it expedient that a believer will be tested. Test is pressing and stressful—one drops faith or grows stronger—when tested. It feels

---

[208] A member of Sunday Assembly, Adelaide. I extend my heartfelt thanks and credits to him for hosting the first ever book launch of this project on July 15, 2018. Post first release of this book on January 15, 2018 two more themes are added to chapter 9 mainly in 9, 9.2 and 9.5.

cold, heartless and uncaring for time being even in faith. The impulse to drop faith can be traced firstly, to the wrong learnings and secondly, to the misunderstanding of philosophy of religion. Some uncalculated and incorrect arguments by the religious authority too compounds the problem before Sceptics. In Quantum physics, navigating away from the naked materialistic view (first view), going by "the Copenhagen" view of interpretation (second view), which makes the Universe a resultant phenomenon by an observer—in line with the prediction of this theory to collapse a wave-function—upon an observer effect. Atheist naturalists ascribe premises to God from a human reference-frame. As a result, they only perceive and measure errors of a human blob. That are obviously deficient from knowing even rudiment wave-function probabilities of an atom. Forgo when these rudiment wave-functions are regulated by freewill such as by human freewill. Besides this, if every observer's perceptions or measurements are considered as real (third view) then MWI (Multi World Interpretation) is accorded. Rather, away from all the above three scenarios if premises are determined by taking in to account predispositions of natural-sciences from God's own view then the truth of apparatus-conditions leading to a divisive observation or measurement of the universe perhaps will be redundant. The Copenhagen interpretation is a false premise used if at all to allude to Creator's plausibility.

The "Schrödinger equation" on the other hand which makes quantum physics greatly usable on a materialistic platform is never deterministic and will only enumerate probabilistic outcomes like that of MWI. Actuality is never captured. This indeterminacy at least with humans who exercise Will at a different level; in quantum mechanical equations, underpins freewill teleology. The fine line of cognisant-magnitude required for a single state of observation is still rudimentary which not only microbes but even viruses in my opinion exercise. Whereas when cognisance is so matured that it no more behaves predictably and is subtle at Will, then, such freewill poses problems to our measurements of it. We are still not seeing God here, in fact, we are unable to see ourselves; except, if we strive for it with great striving which qualifies humans for who we are. This can at most be our perception of premises to comprehend reality. Within our own domain of need we see our math fails us to describe everything we set out to measure.

Clearly, with these equations our claims to measure God is invalid. Quote: *"Certainly, science can never be absolutely certain that it has arrived at a final and complete description of the physical world; so, it will always remain a possibility that present quantum mechanics is incomplete and that a modification of its formalism will resolve all its puzzles, paradoxes, and conundrums."*[209]

The question we need to ask is can we really explain everything without 'God' being in the equation? Admittingly, we cannot even explain event-time of matter. If we could we would truly begin our first step in eliminating the need for God. For example, we can't measure when the Carbon 14 decays to Nitrogen 14, but we only know the probabilistic decay curve. If we knew the time of decay of every nucleus then we could channel packets of energy to prevent the decay. Inducing decay as in a chemical reaction of disposable nuclear physics is a different story. Our present ability is we realise events after they happen. We cannot pre-emptively act due to lack of knowledge owing to many uncertainties. This is where time comes in to play with God. For any worth making the Copenhagen interpretation a premise for God-figure observer to realise the Universe as it is now in the making is a false premise. Rather God is the creator of the Universe and not an observer as -we- are facing nature unfold. And He plays it all in a function of time. Quran 7:185 *"Do they not look into the realm of the universe and the earth and everything that Allah has created and [think] that perhaps their time also nears? So, in what statement after this will they believe?"* But the unique solutions and definite outcomes instead of the never-ending strings of looming possibilities is evidence of a Creator forcing one solution to unveil a controlled Cosmic reality rather than MWI (Many Worlds Interpretation). In this God informs us of the future certainty to unfold—we call it a prediction in our science—He calls it a definite set of events to unfold; which we can today verify. Unlike what we struggle with determining the ends of probabilities in arriving at the states of a system using the quantum mechanical equations.

---

[209] Does Quantum Physics make it easier to believe in God? By Stephen M. Barr, Professor of Physics at the University of Delaware published on July 10, 2012 on "bigquestionsonline.com."

The agnostic world is inundated with premises concerning God that are blatantly incorrect such as:[210]

1. *In classical theism, God does not make measurements. His knowledge is not "posteriori" but "a priori." That is, He does not learn about things as a result of their happening, but rather they happen because he knows them, so speak-* Is false. Because, Quran 45:23 *"Have you seen he who has taken as his god his (own) desire, and Allah has sent him astray* **upon knowledge***..."* In our domain where wave-functions are governed by freewill even a thought conceived is instantly brought to God's awareness. And He becomes cognizant of it **by knowledge.** God becomes knowledgeable as opposed to the assumed false premise.

2. *The wavefunction encodes what some observer is in a position to assert about the physical world given some knowledge that he already possesses by virtue of other measurements or observations made by him. So, the observer is, practically by definition, some being who acquires information by means of physically interacting with the rest of the universe. God, however, (according to traditional theism) does not physically interact with things in the universe, as he is not a physical entity-* Is false. Because, Quran 6:103 *"No vision grasps Him, but He fully encompasses their vision and He is subtle and acquainted"* and Quran 6:101 *"...He created everything and keeps knowledge of it"*, these verses award God unique premises of a Creator of things in a pre-determined and a measured way and then who traces them continually unlike the 'non-interacting God' a false premise understood of traditional theism. The tracing is of a different level where God comes to see how little we see. That we cannot see the wave-function until it collapses. But God is above such a limitation and while under God the intactness of a wave-function is **tenable.**

3. *Accounting for God's omniscience, all seeing and all knowledgeableness analogous to human frame of abilities-* Is false. Because, this is where Philosophers/Scientists concede "God in the image of a man", quite

---

[210] Using Quantum Physics to "Prove" God's Existence by Andrew Zimmerman Jones updated April 01, 2018 on "thoughtco.com."

literally. In fact, this phrase is not part of the words communicated by God in scripture. It is an addition in religious literature, as such, no problems in using this metaphor. God's way of observing us now or previously will not amount to -no observer- effect. Instead this is the expected result from us to demonstrate pure faith that we have to conduct as if under the watchful eyes of God. But freewill earns God's forgiveness in the greater Wisdom of God. Yes, if God were to show up as when in Judgement day the drama would end and our freewill would collapse causing an observer effect- but for the teleology of a test to continue God for now is the Unseen. Until Judgement day this matter is clearly deferred. Sam Harris argues freewill incompatibility, he has this to say: "Based on future neuroscience mapping of brain configurations of individuals for rogue state of affairs measurements in a scientific way can identify impending criminal behaviours and therefore we can inhibit them on a priori to committing a crime." Which according to him saves society harm and judicial time in processing criminal investigations. Heavens! How close do our liberties and freedoms inch to be threatened—by irrational endeavours—heedlessness to God endangers human potential; despite its claims of civility.

Clearly, the realm of Unseen is a reality before our vision. Almighty God is not the entity that we invoke and refer to as 'god' of the gaps. Our knowledge of everything or whatever does not discredit God's existence. Scientific research need not prove God's existence. If it does, then it will be unjust of God on those disbelievers of old who unfairly couldn't benefit from science to belief in God. God's being is fundamental to human consciousness—not needing—revelation by God. And so is presenting for Judgement before Him and acting righteously for the love of Him and then desisting from wrong for the fear of Him. The natural question one may then ask is: Why do we need scripture? And the answer is to know the premises; otherwise we may pose an excuse before God.

> Quran 6:157. Or lest you should say: "If the Book had only been sent down to us, we should have followed its guidance better than they." Now so it has come your way from your Lord- clarifyingly and a guide and a mercy...

Understanding doomsday unfold is by far the easiest subject in the Quran. Having said the science of it is the most complex subject an indeterminate for knowing its time of happening. There are a few indeterminates which only God alone keeps the knowledge thereof: *Quran 31:34 "Indeed, Allah (alone) has knowledge of the Hour (Doomsday) and sends down the rain and knows what is in the wombs. And no soul perceives what it will earn tomorrow, and no soul perceives in what land it will die. Indeed, Allah is Knowing and Acquainted."* Among the virtual realities listed in this verse here, rain is close to be known yet uncertain. Our most stately weather models can but only forecast. The uncertainty principle, also called Heisenberg uncertainty principle, articulated (1927) by the German physicist Werner Heisenberg,[211] says that the position and the velocity of an object cannot both be measured exactly, at the same time, even in theory. Apart from the problem of quantum mechanics in its study of dynamics of electrons in an atomic model, macroscale events such as expectancy of rain also poses similar uncertainties. All we could do is forecast the weather despite numerous mishaps. It may rain or perhaps may not, time and place of rain is still very elusive. Further, according to the verse God alone is aware of the developing creature in the womb. As God becomes aware of the developing creature, He lays down tests for it for its unique abilities in exercising freewill. Not everyone goes that extra mile in everything: *Quran 2:286 "On no soul does Allah place a burden greater than it can bear. It gets every good that it earns, and it suffers every ill that it earns…"* Having said, doomsday is closely connected to human responses to God on -belief and righteous conduct- (BRC). God will delay the doomsday from happening until BRC uncertainty-wave collapses: *Quran 11:104 "Nor shall We (Allah) delay it but for a term appointed."* When God becomes aware of our dissensions for BRC despite the prosperous health of any number of select societies in the wake of prevailing greater fasad (mischievousness); God will determine to conclude the doomsday. There is a stretch but when it expires in the context of BRC whose brink God alone fathoms, the end will brook no delay. This is because God is the only provider of faith and the faithfuls can only help spread the word to conclude God's argument for final determinism.[212]

---

[211] Niels Bohr and Werner Heisenberg in the years 1925 to 1927 largely devised the Copenhagen interpretation.

[212] Determinism: Also a decree of God to interfere into human affairs at the level of wider population. It mostly concludes in death & destruction and can also result in suffering and losses.

The Quran reveals various sequences of doomsday in detail. But how distant reaching are the effects of doomsday into Cosmos? Concerning this subject, the Quran localises doomsday phenomenon to occur within the near celestial pockets of Cosmos. The Quran is very suggestive of dooming solar neighbourhood in its entirety. And then our translocation to another space within regions of Milky Way is a plausibility. A home in another constellation in the sense of a new neighbourhood is credible. The present solar neighbourhood in Orion house could be obliterated by gravitating it toward the galactic centre. A region where remnants of stars are expended in the humongous sink known as a blackhole transforming into a Quasar.

An alternate take or a still plausible perspective of doom is that the entire galaxy is set into commotion by the drawing away of dark-matter from about the silhouettes of the Milky Way triggering the full galaxy to collapse entirely instead of the pocketed implosion of dark-matter (the heavy substance permeating the fabric of space) from within the interstellar regions of our solar neighbourhood. In such a scenario given to unfold for doomsday we would be translocated to a different galaxy altogether to commence the promised afterlife.

As seen earlier in section 6.7 dark matter and energy bears the heavy impetus to cause a sudden doom. The elusive dark matter sky when locally displaced collapses the encompassed region within it. The Quran is very indicative of suggesting this that the realm of our doom is a local phenomenon in the wider Cosmos. Traditional misunderstanding of the whole of universe experiencing a big crunch is invalid as studies reveal startling data from the Quran's purview. Assertively, a reverse crunch of universe's expansion theory isn't implied in the Quran either. Instead expressly the Quran is found to reveal a very local phenomenon of the Cosmic doom of our world.

9.1 QUANTUM STATES AND RECORDING

Allah is unique from His creation. Not even Angels comprehend of knowledge except as God wills for them: *Quran 25:65 "Say: None in the skies or on earth, except Allah, knows what is hidden: nor can they perceive when they shall be raised up (for Judgment)."* There is a clear limitation as to how much Angels are given to know. One among the unknowns is the time of doomsday. Angels do have obvious limitations to perceive full knowledge of what the earth in entirety contains. Almighty is dominant over His creations. His knowledge and reach permeate every sphere. He is the most immanent الباطن (al-batin): *Quran 6:59 "With Him are the keys of the unseen, the treasures that none knows but He. He knows whatever there is on the earth and in the sea. Not a leaf does fall but is in His knowledge: there is not a grain in the darkness (or depths) of the earth, nor anything fresh or dry (green or withered) but is (inscribed) in a clear record."* This verse tells of Almighty's nature of awareness keeping imbedded in the Cosmos or the incomprehensible nature which permeates entirety. Often confused the human mind tricks to concede God's presence literally in everything. As Hindus circumambulate various objects and substances.[213] Waving aside the assertion of God permeating materials; this falsity cannot sanction worship of them. Instead even a leaf withering is in the awareness of God. The many counts of grains of sand or things in the depths of ocean or within earth and things fresh or rotting all energy levels of all things at their quantum states in the realm perceivable to us of baryonic material are in the firm awareness of Lord. Besides things are recorded and everything is enumerated: *Quran 22:70 "Don't you know that Allah knows all that is in skies and on earth? Indeed, it is all in a Record, and that is easy for Allah."*

With mankind it is even more serious. As we are given to arbitration God prepares things for this very purpose: *Quran 36:12 "Verily We (Allah) shall give life to the dead, and We record that which they send before and that which they leave behind, and of all things have We taken account in a clear Book (of evidence)."* Allah's knowledge is not disputed but mankind would seek evidences from witnesses therefore God makes Angels do this work diligently: *Quran 43:80 "Or do they think that We (Allah) hear not their secrets and their private counsels? Indeed (We do),*

---

[213] Various objects besides idols and temples, trees, sanctums of water ponds, ceremonial fires etc.

and Our messengers are by them (entangled) to record." And *Quran 54:52-53 "All that they do is noted in (their) Books (of Deeds): Every matter, small and great, is on record."*

> Quran 10:61 And in whatever engagement you may be, and whichever portion you may recite from the Quran- and whatever deed you (mankind) may be performing, We (Allah) are witnesses thereof, as you are deeply engrossed therein. Nor is absent from your Lord (so much as) the mass of an atom in the earth or in heaven **or (things) smaller than it** or greater but that it is in a clear register.

Science of Quantum mechanics has been enormously successful in giving correct results in situations to which it has been applied. [214] However, there is an intriguing paradox. The foundations of the subject contain unresolved problems. Particularly problems concerning the nature of measurement.[215] Due to which uncertainty or indeterminism is realised. Modern science has identified that atoms do not represent the smallest unit of matter anymore. Particles called quarks and leptons seem to be the fundamental building blocks. But physicists are still far from understanding why a proton has about 2000 times more mass than an electron and they are on the look for even smaller particles. People in seventh century Arabia hardly fathomed atoms or even particles smaller than it. Whatever we humans are involved in God bears first testimony of it while we are fully engrossed in it. He guides us for what is allowed for us to gather knowledge needed for building our civilization. Even things or particles smaller than atoms are clearly in a register. Further all creatures are enumerated: Quran *11:6 "There is no moving creature on earth, but its sustenance depends on Allah: He knows where it dwells and its place of deposit: All is in a clear Record."* Every atom has a certain allocation of energy. These states are called quantized energy states. Per the decree of God every atom of the non-lively is regulated in rigid mathematical formulations and of the lively is regulated in subtle math-formulations of consciousness. From recipients of freewill it is required they justify for energy level fluctuations to doubt or disbelieve in Almighty. Also, from normalness towards extreme behaviour which potentially hurts or infringes into the

---

[214] Title search: What Has Quantum Mechanics Ever Done for Us? by Chad Orzel on forbes.com.
[215] Title search: Quantum mechanics by Gordon Leslie Squires on britannica.com.

perimeters of others. Therefore, all of this is on a continual recording. Literally everything unfolding in the realm of this planet, then every thought consciously conceived and every word uttered and further every subconscious wandering are in the awareness of Allah: *Quran 50:16 "It was We Who created man, and We know what dark suggestions his soul makes to him: for We are nearer to him than (his) jugular vein."* And thus, all states of our mind are considered if not for recording but in God's awareness. Owing to the rule of Mercy with Allah not every wrong thought progresses to become a sin but strong intentions driving those thoughts and lack of seeking forgiveness for them does graduate to call us to account for it.

Naturalists to eclipse charm off the religious say: simple as washing hands before meals isn't even prescribed in religion, alluding to the monstrous world of the microbes beneath. Not knowing much of the eastern tradition, the westerner naturalists is at fault here. It is so fundamental a practise to cleanse one's hands with water before meals that it is not a necessary prescription required in a scripture. In fact, in addition to a clear list of mandates of personal hygiene, it is well known of Prophet Muhammad counselling for keeping calm energy levels. When asked the question, who makes a strong person, in response to 'he who can wrestle people down', the Prophet educated his pupil that 'it is he who controls himself when in anger.' It is worth reflecting on the sort of reflections were set into Prophet Muhammad by the Quran to say about the balanced life he advised us to be on. For 'anger hurts a person's faith and religion' he said. God guided Muhammad by His revelations on fundamental sciences, such as told in: *Quran 42:37 "And those who avoid the major sins and immoralities, and when they are angry, they forgive."* Thus, working towards nurturing calmness and seeking truth is as essential to neutralise one's state of living to being natural towards uprightness in faith. The Quran teaches that lying is one of the big sins, hypocrisy is one of the most condemned states of being and fraud is one of the worst forms of deception. It teaches that speaking truth and standing up for it are qualities possessed by the righteous. Patience, perseverance and constancy are necessary virtues and so is taking care of the weaker members of the society such as orphans, widows and servants (equalling to employees' rights). The Prophet said: 'I and the care taker of the orphan will be neighbours in paradise' and kindness and compassion must be shared with the non-human world. Not getting angry and being away from sadness is told by Prophet as freeing oneself from Satan. Such are the characteristics that must go into estimating the measure of a moderate and a truly flourishing society. These tokens are clearly not just to save oneself from the world of deadly microbes but to emancipate oneself from the dreadfulness of a hell.

9.2 THE LAW OF DECREE

Quran 97:4 "Therein (in a night in Ramadhan, an Islamic holy month of fasting) come down the Angels and the Spirit (Gabriel) by Allah's permission, on every Errand." Every lunar year in a special earmarked night. All matters of births, deaths, sustenance, determinisms of the trials to examine faith and of type- justifications for disbelief & sins are decreed for the coming annual season. As matters are given to unfold upon individuals or upon a community, all events thus are scaled, determined and mapped on site earth for which numerous Angels spearheaded by Gabriel descend in a night. These are extraordinary mathematics readied for manifestations: *Quran 57:22 "No **calamity** strikes on earth or any **mishap** befalls you but is prior determined in a book before We (Allah) bring it into existence. That is truly easy for Allah."* The Arabic word here مصيبة (museebah) which means misfortune, calamity, disaster, affliction, adversity etc. is a sophisticated decree of God. However, traditional understanding of this concept has been very rudimentary. Due to the Islamic scholarships' inability to increase reflections over this subject of great importance as very early on anarchy troubled the Muslim world, since then purist scholarships operating under opportunist rulers, had invented a recipe to monopolize over the legitimate voices from making true progress upon the teachings of the Quran. As always, the first right opinion from the Quran surfaces in favour of public awareness arousing them to partake in civil affairs, which has been perceived as a great threat by every ingrate King. Pre-emptive of this, right opinions were systemically delinked, and truth degraded. Up until today, Muslim community is often endemic in disqualifying critical thought; even a step towards understanding the Quran to them poses a significant violation of status-quo hierarchy of staged scholarships.

> Quran 42:30 And whatever strikes you of disaster- it is for what your hands have earned; but He pardons much.

The Prophet's crystal clear dictum: 'A faithful will not err the same again', is based on the principle communicated in a couple of verses as follows: Quran *3:135 "And those who, when they commit an immorality or wrong themselves (by transgression), remember Allah and seek forgiveness for their sins- and who can forgive sins except Allah?- and (to those who) do not persist in what they have done while they know"* and Quran *3:139 "So do not weaken and do not grieve, and you will gain mastery if you are (true) believers."*

Determinisms in the Quran could materialise in various ways in one's life. They are on a normative level tied to our use of intellect. Say, when in dealing business if you conceded loses from an untrustworthy vendor then not ensuring to invoke the procedures of law and indemnity to ratify the next deal would potentially make you vulnerable for yet another loss and will also make you a jerky person. The Quran having given guidance to avoid such a misfortune مصيبة (museebah) due to imprudence which is a very superficial understanding of this concept; it tells us of the clear trials from God. If we put our intellect to use God owes us success lest there is loss! Which is quite different to a case where God is intending to test our faith in Him. Despite our best attempts one may lose gains in market or fail to meet the ends but nothing of this must prevent us from being true believers and righteous men & women. The idea among scholarships today that **every individual is predetermined an inmate to hell or heaven** is not a Quranic premise and even its roots are not found in traditional Islam. Due to many of our shortcomings and various weaknesses God pardons much wrongful thought and behaviour from us; but yes, if we persevere to better our lives by acting rightly (if needed by way of sacrifice) then God is truly appraising of our striving towards determining truer positions: Quran *5:39 "But whoever repents after his wrongdoing and reforms, indeed, Allah will turn to him in forgiveness. Indeed, Allah is Forgiving and Merciful."*

When we are self-aware, we can alter misplaced emotions because we control the thoughts that cause them. Self-observing profoundly changes the way our brain works. It activates subtle regions in the brain that gives us an incredible amount of control over our feelings.[216] If we are just denying truth for the sake of denying, we are surrounded with ignorance due to determinisms of localised thought processes and thus freedom or freewill is lost. Again, by Jesus's popular quote: 'seek ye the truth and the truth shall set you free.' Freewill is a responsibility, the ability of it to acquire truth via pursuit of knowledge can't be wasted. The non-human world apparently operates deterministically except by Allah's will and interjection, but with humans it is not at all rudimentary akin to animals. We are prevented from being reflexive nor are we overly docile beings. In our sophistication we have acclaimed superior-will which is freewill and thus we

---

[216] Title search: Awesome Human Brain and Quantum Physics Documentary 2016 HD, published by Albert Embrey on 29 May 2016.

reason, adjudicate and justify. It is hoped of this work that it will help penetrate the shell of closed views of staunch dissidents such as Sam Harris and Richard Dawkins and provide them with another perspective to reform their standpoints.

Eventual time of doomsday is an unknown. Astronomy from the depths of unknowns rises to unwind a number estimating sun to die and scornfully lets the religious know to relax for at least a few billion more years to pass. The Quran doesn't secede from real conclusions of science; but it fore informs by superior knowledge. As seen in section 6.7 doomsday is awaited to shudder in from an elusive dimension. God's Will is covered by His mechanisms in place to bring it to pass in our age or prolong it's happening precisely where karma determined by the BRC index justifies it. If we could hearken to correct belief and righteousness then Allah will prolong it's unfolding until for a time when we will again be in rebellion and overload the world with many injustices. By our present wrong doings, we're on a haste race. The able Western world not heeding to righteousness in the best interests of humanity is being caught red-handed. It fortifies staunch oppressors in middle-east such as with the state of Saudis and overthrows others to keep geostrategic interests even by forging lies to the extent of invoking God's holy name for invasions. When strategic interests for the West are deemed non-lucrative, principles of freedom, liberty and human rights are buried in dungeons. What results is violations, like in Palestinian territory and flaring new uprisings with vigour and zeal like in Syria. If this is the response of a civilisation, surely determinisms inch near. In the end when God deems humanity as a whole is not for reform, then judgement becomes imminent: *Quran 34:3 "But those who disbelieve say, "The Hour will not come to us." Say, "Yes, by my Lord, it will surely come to you. (Allah is) the Knower of the unseen." Not absent from Him is an atom's weight within the skies or within the earth or (what is) smaller than that or greater, except that it is in a clear register."*

> Quran 72:25-26 Say (O Muhammad), "If I were to know what you are promised is near or if my Lord will grant for it a (long) period." Knower of the unseen He does not disclose the unseen to anyone.

> 27-28. Except whom He has approved of (Angel) messengers, and indeed, He sends before each messenger and behind him safe-keepers. That He may know that they have conveyed the

messages of their Lord; and He has encompassed whatever is with them and has 'enumerated' all things in **number**.

To those Angels posted in Laniakea supercluster, they receive codecs as said in the verse 72:27-28. Even Angels do not encompass knowledge of the Lord and of the unseen except what is given for timely action. And all things reduce to witnessing, even the sphere surrounding Angel's work. Allah has in His law designated numerals to everything. And all states of this transmission process to Angels in this zone are encompassed. Within human domain our thoughts, intentions, words, actions and responses to God's communications all quantised states are enumerated in numerals.

> Quran 35:11 Allah did create you from dust, then from a sperm-drop; then He made you mates. And no female conceives, nor does she give birth except with His knowledge. And no aged person is granted (additional) life nor is his lifespan lessened but that it is in a register. Indeed, that for Allah is easy.

The decree of God in human spear works on supplications (Arabic: Dua) which is to seek forgiveness of and ask from Allah. The knowing and deciding God if He so wishes, will accept our dua (prayer) and will change the decree to favour us. This works in individual capacity and also becomes enacted by a community. Nothing of what happens in this world is omitted but is recorded in the Book.[217]

Failing to believe or working goodness in a given timeframe brings us close to the threat of facing determinisms. If mankind does not realise and be subservient to God then troubles lurk near: Quran *17:58 "There is not a population, but We shall destroy it before the Day of Judgment or punish it with a dreadful Penalty: that is written in the (eternal) Record."* Thus, it is time for humanity to shun the wrong ways of our predecessors and amend for the betterment of our global civilisation. We do not have to worry about our predecessors' status with God, for He is keeping account of them: *Quran 50:4 "We (Allah) already know how much of them the earth takes away: With Us is a record guarding (the full account)."* It is correct to worry about ourselves more than lamenting on past tense. But some pseudo

---

[217] Quran 6:38 There is not an animal (that lives) on the earth, nor a bird that flies on its wings, but (forms part of) communities like you. Nothing have we omitted from the Book, and they (all) shall be gathered to their Lord in the end.

heralders of science with naive sense of absoluteness are taking to stage to brainwash scores with delusions. Not realising that the indeterminate of doomsday or other retributions of God coming our way are never given to us in advance for us to perceive. They wait to be simply overwhelmed when it is too late. Clearly with God rests the eventuality.

When determinisms from Allah approach such as coming of a messenger, it means eventualities are very close: *Quran 8:13 "…if any contend against Allah and His Messenger, Allah is strict in punishment."* Here again there are two types. A Prophet (Arabic: nabi) and a Messenger (Arabic: rasul) have paired but ranged connotations. Such as any holding the divine office by nature is a Prophet with a level of exception. A Prophet means he is a news giver, a reformer and a messenger such as Zachariah (Arabic: Zakariya) and John (Arabic: Yahya) who are essentially prophecy givers and reformers; here a messenger simply means 'sent by God'. And on the other end of the spectrum there comes a declared **Messenger** sent to source a revolution for example Noah, Abraham, Moses, Jesus and Muhammad bringing clear warnings of God to their designated people to be obeyed or be engulfed in wrath. Moses was a Messenger to Pharaoh and a reformer to the house of Israel. Jesus was sent as a Messenger to the house of Israel and a reformer also. Noah and Muhammad respectively were Messengers first and also news givers and reformers. Lot (Arabic: Lut) to his folks, is among the exceptions joining the line of **Messengers** who are not news givers (Prophets: nabi) or revolutionary in the sense of their mission, neither do they receipt covenants such as Torah or Psalms of David but in essence they deal with an identified error in societies that they are sent to correct. Such a line of Messengers was sent to- the people of Palmyra, the city of Petra and others with Lot being the last in that category. Unlike this line who are clearly appointed by divine commission there are lastly, self-conscious believers acting as **messengers** among commons who do the role of a Messenger-ship without having proclaimed divinely by God. When these perform on the level which qualifies to invoke determinisms of God then the law still comes to effect, such as those who scaled up the mission to invoke determinism of God upon Pompeii after the period of divine commission to Jesus Christ. Such people today proclaiming God's holy name are the bearers of the same warnings from God.

Therefore, in this analysis you see any who endeavours in any of the divinely appointed Messenger's (rasul) footsteps reminding people of God are aided by

Him. Also, as the target societies cannot overcome the Messengers as in the example of Jesus Christ whom the disbelieving Jews could not kill,[218] this decree of Allah then becomes effective upon those who transgress the warnings of God given to them by reformer messengers. Making them lowly as God sends upon them natural calamities or armies of numerous men as in the example of armies of disbelievers[219] (approved of God) decimating the house of Israel to justify for their disobedience to God.

On the other hand the reformers (nabi) without the commissioning of Messenger-ship can get killed like the house of Israel killing many sent by God such as Zachariah and John: *Quran 2:87 "And We did certainly give Moses the Torah and followed up after him with messengers. And We gave Jesus, the son of Mary, clear proofs and supported him with the <u>Holy Spirit</u>. But is it (not) that every time a messenger came to you, (O Children of Israel), with what your souls did not desire, you were arrogant? And a party (of messengers) you denied and another party <u>you killed</u>."* [220]

---

[218] Per Quran Christ was raised to God. Quran 4:157 And (for) their saying, "Indeed, we have killed the Messiah, Jesus, the son of Mary, the Messenger of Allah." But they did not kill him, nor did they crucify him; (it) was made to appear such. And indeed, those who differ over it are in doubt about it. They have no knowledge of it except the following of assumption. And they did not kill him, for certain.

[219] House of Israel was quelled twice once by Persians in 589 BC and in 70 A.D. by Romans.

[220] In the religious legacy of Indus valley *Avatar* is a revered soul representing God's mission—a chosen one by God of the past for every age. Modern Hindus perhaps will disagree with this elucidation. Because of the contemporary baggage of lensing this concept via literal studies of God incarnates. An attempt to go inclusive away from the premises of Sanatana dharma of the Vedic.

Whenever decay of religion happens and rise of unrighteousness prevails the symbolic telling was that I God will manifest to protect the good, for the destruction of evil and further to establish righteousness, I will be born (symbolic of a Prophet) in every age. References: Bhagavad Gita 4:7-8 and Bhagavata Purana khand 8 Adhiyay 24 Shloka 56. The point of information is this concept was coined post Vedas. The manifestation of *MahaRishis* as per Vedas if translatable as Prophets, then *Avatar* is the post extolling of this concept in the praise and symbolic honouring of God's missions of the past. Much later these high emotions resulted in apocrypha of God incarnates and thus anthropomorphism was deeply rooted among the teachings that today *Avatar* is extremely proliferated and has been even applied to non-human

In yet another decree of God in absence of any warners making an argument on God's behalf, the measure is then by the level of oppression that societies come to display because God is ever the listener of the supplications and calls of the oppressed.

> Quran 74:32-34 Nay– By the Moon! Behold at Night when it retreats. And at Dawn it shines.

> 35. It is one of the momentous portents.

> 36-38. A warning to human being. Let those of you who prefer, step up or lag behind. Every soul will be held hostage for its actions.

Quran verses 74:32-34 have been traditionally singly interpreted. Which means each phenomenon i.e. the Moon, the Night and the Dawn (not Day) are

---

beings. And that *Rishis* today is very commonly applied to religious peers & seers in Hinduism, its reservation to Prophets of God by God's commissioning has been forgotten and faded.

A similar persuasion has soiled Buddhism to the degree of dismissing God entirely— where the ontology of God as the final Judge is doomed. A proliferation of this nature, disrupting the existence of God and the office of Prophethood simply falls apart upon logical & scriptural inquiry. This is common of humans as we get to see in case of Jesus too, many systemic attempts have been made to make him the divine. This is where the jinns are at their best to delude us. They make us think of this charming unification that such coagulation is a serene pleasure of spirituality. In fact, this leads to the deconstruction of the religion of God and His laws of merciful communications with mankind for guidance which are thus undermined.

Prophet Muhammad's traditions speak of more than one hundred thousand God's commissions upon earth involving communities of various nations of which the prophecy was sealed with him in Arabia being the final in the tradition of God's communication to mankind. It is therefore plausible that many personalities of Indian fame such as Rama could be deemed a Prophet of God in the line of Noah (Manu) or otherwise a descendant from companion believers of Noah. Such a tradition of prophecy among Jains of India is revered as the *tirthankaras*. From South America to Africa and of the China & Asia east, all nations in the past from Mongols to Vietnamese to Indonesians whenever they organised societies based on justices they came to have receipted a Prophet from God. As God says: Quran 16:36 *And We certainly sent into every nation a messenger, [saying], "Worship Allah and avoid blameworthy innovations."*

independently meant to be alluding to a sign. But in this passage verse 74:35 forces a correction.

'Blood Moon' is a very modern descriptor[221] still used after Hagee's prophecy proved to be hokum, see **Figure 37**. The label lives on in a secular way to fancy this event to popularity. As the subconscious psych would readily concede to anything of fright in the sky as it did in human history.

> Quran 84:16-18 I (God) swear by the twilight.[222] By the night and what it shrouds.[223] By the Moon as it collects.[224]
>
> 19-20. You will progress from stage to stage. So, what is the matter they do not believe?

I conveniently missed out on commenting over these two different passages in the first release of this book, self-published on Jan 15, 2018.

Jan 30, 2018 saw one of its kind lunar event. It was a rare super blue blood moon. This Moon was cited on Wednesday morning hours in the USA and set behind the city of Jerusalem in the evening hours on Jan 31, 2018.[225] Since my book became self-published, I have continued to update it. After the super blue blood moon event; still having not taken notice of the above passages in the Quran, I

---

[221] This is in part due to the release of a book called Four Blood Moons by Christian minister John Hagee highlighting a lunar sequence of four total eclipses that occurred between 2014 and 2015. Source: http://observer.com/2018/07/blood-moon-lunar-eclipse-history/.

[222] The Arabic form بالشفق (bi-shafaq) used here is the main object that combines preposition 'bi' with the noun 'shafaq' to mean 'with the twilight'. This object opens the context of this discrete passage.

[223] The word وسق (wasaq) means 'to gather or collect'. The Night umbra of earth collects the moon within.

[224] Again, the word تسق (tasaq) is just another form of وسق (wasaq). Here the moon is referred to collect red-light from the twilight zone. Thus, the blood moon appearance.

[225] A blue moon (a second full moon in a calendar month), a super moon (when the moon is unusually close to Earth, making it bigger and brighter) and a blood moon (a moment during an eclipse when the moon appears red) will all coincide for the first time since 1866.

got endorsed into my next work on explaining creation details of earth and origins of life. Until when I read and wondered on these two passages of the Quran 74:32-38 and 84:16-20 way down in July 2018 just before the second blood moon event in the year. I thought of including this essay here.

In the wake of positive exchanges with members of Sunday Assembly, Adelaide at my first book launch. I ended up including an essay on Quantum physics as seen in section 9. And an essay on blood moon as the need for it coincided. Yes, with sceptics' assembly of which Peter Morris is a member was my first book launch. I thought University presses; Book publishers and Book stores less represent the civility of a "Western" society when they cited this book as not suitable for a "Western" market place. I have to admit, even the Muslim response has been hugely discouraging. Out of the many Muslim organisations worldwide I reached out, all most all, were disinterested. Lacking the tenacity to read any of this many among the Muslim community who know me or whom I know in Adelaide despite having lived in the "Western" world where critical thought is promoted as part of the value system are either silent or chose to deliberately ignore this contribution. Prejudiced they turn away. Admittingly because of the self-critiquing narrative brought forward in this book which may be to their dislike. I wonder at this point, if all of the human tribes presently living are innocent only deserving praises and not rebuke. I wonder if this be the case then why are the world affairs at its worst? Clearly the "Western" world is a better place on earth today to live than any of the Muslim lands. One of the reasons why Muslims from all ethnic backgrounds flock to these countries yet reluctant on critiquing their wrongs. Individuals form families and become nations. It is important that individuals learn their weaknesses and to pick upon their own errors. This will only make them better. If any deems on becoming an Angle, then perhaps wait for what will face you of the wrath of God. Common Westerners have much appreciated this book without any hiccups. After this small success and positive discussions with the readers of this book who thought that the scientific community would be able to verify claims of the Quran presented herein; I'm hopeful of a day when it will be tackled by the elite. Along with this new inclusion, I have reworded some portions of the manuscript for better comprehension.

On July 25, 2018 second time in the year blood moon was on the news again. It prompted me to look into verses which I had read a few days earlier. On the first eve on Jan 30, 2018 I did not wonder on this subject as I had conveniently missed these verses from considering into my book. The event hit the news not until early July, by my search for first information on this eve.

An animation on 'timeanddate.com' showed approximate times for the eclipse in New Delhi. Stages and times of the eclipse marked the Penumbral Eclipse begins at 10:44 pm Fri, 27 July and Penumbral Eclipse ends at 4:58 am Sat, 28 July, the total event to last about 6 hours. With Maximum Eclipse of Moon closest to the centre of Earth's shadow at 1.51 am Sat, 28 July. All times reported here are local time (IST) for New Delhi.[226]

I have to admit that interpretation of scriptural text unfolds in this fashion. While the Sceptics would remark this way of understanding or interpreting ancient texts as selectively cherry picking and interpreting as per convenience. Fortunately, this is not the case. Most of the book's contents have been derived on available knowledge then predicated on this knowledge numerous postulates have been enumerated on the authority of the scripture alone. Days earlier to reading news on July's blood moon I wondered on these verses without any major conclusions. Especially on the astronomical front of celestial mechanics. I have to admit I had difficulty imagining what sort of astronomical arrangement the Quran was alluding to as the major portent. Except that I knew at first instance that it wasn't surely how traditional reading of these passages had been thought.

Quran verses 74:32-34 have been traditionally singly interpreted. Which means each phenomenon i.e. the Moon, the Night and the Dawn are independently meant to be a sign. However, I knew under the guidance of verse 74:35 that this whole passage elates as one single mighty portent. Blood Moon, a phenomenon which recourses in celestial sequences of full moons i.e. total lunar eclipses; on an average at least 60-80 full eclipses concur in a given century. Of those at least a few last over 100 mins in the umbra counted as maximum durations in a given century. This occurrence is simply a celestial reference that God uses to warn mankind: Quran 74:36-37 *"A warning to human being. Let those of you who prefer, step up or lag behind."* These recurring astronomical sequences of the moon are stage completions of a celestial timeline just like the new moon showing up every lunar month. The full moons that hung in the umbra (centre of Earth's shadow) for long periods of over 100 mins of time recur as the Quran clarifies that the

---

[226] https://www.timeanddate.com/eclipse/in/india/new-delhi.

penumbral retreat happens at night and lasts until dawn. By far, it is expected even by the society of Earthsky, that July 27, 2018 is the longest of retreats in 21st century. Analysis of the Quran's passage is as follows:

In his early call to Islam the Prophet was a bit timid after left estranged to stand up to the prejudices and biases of his people. Then God reverbs infusing strength in him asking the Prophet to stand and Warn his fellow men.

> Quran 74:1-3 You (O' Muhammad), wrapped in your cloak. Arise and give Warning. By proclaiming the greatness of your Lord.

The surah begins asking the Prophet to Warn mankind. The revelation of this bit from the surah about first 10 verses were in the earliest phase of call to Islam. Later verses of this surah followed when a band of Muslim were led into first immigration (self-exile) to Abyssinia to avoid persecution in Makkah. Verses 74:11-31 were directed to the disbeliever Military chief Walid ibn al-Mughirah al-Makhzumi, the chief of the Banu Makhzum clan of the Quraysh tribe. His clan was responsible for warfare and security of the city. Walid ibn al-Walid and Khalid ibn al-Walid were two influential sons of -al-Makhzumi- who both embraced Islam. Former was an early convert and the later was a later convert; but who led victorious forces of Islam upon the Persians and Roman arrogance for their reconnaissance and killing of Bedouin Arab converts to Islam in the border regions. While -al-Makhzumi- himself earned the wrath of God for long-time plotting against Islam. Khalid at his conversion began learning the Quran with this surah which rebuked and condemned his father to hell, saying: "I wish to overcome the love of my father by repeating after this Quran." Such was the transformation of seekers of this faith that they culled their prejudices and biases from within its roots. In this very theme a verse after concluding doom on Walid. The Quran announced a momentous portent. Nay, by the Moon it says:

> Quran 74:32-38 Nay– By the **Moon**!
>
> 33. Behold at Night when it **retreats**.
>
> 34. And at Dawn it shines.
>
> 35. It is one of the momentous portents.
>
> 36. A warning to human being.

37. Let those of you who prefer, step up or lag behind.

38. Every soul will be held hostage for its actions.

In verse 33 the Arabic word dubur (backside) translated as '**retreat**' is used to allude to an astronomical sequence. This verse should not be interpreted singly making the 'Night' as the object. As has been traditionally translated to infer: 'By the Night as it departs'. Instead the object is identified at the beginning of the context which in this case is the '**Moon**'. Similar contexts in verses 81:17-18, 89:4 and 91:4 for the oaths by the 'Night' also contextualise 'By the Night' as subject only during which an event is alluded to occur of the 'direct object' first invoked in the context. The word dubur (backside) therefore means while the Moon faces the sun to shine its light as seen in chapter 2.2 it is retreating into the backyard of the Earth's 'Night' where it faces away from shining the direct sunlight from its home in the sky space. This backyard is the penumbra, dark side of the Earth, see **Figure 37**.

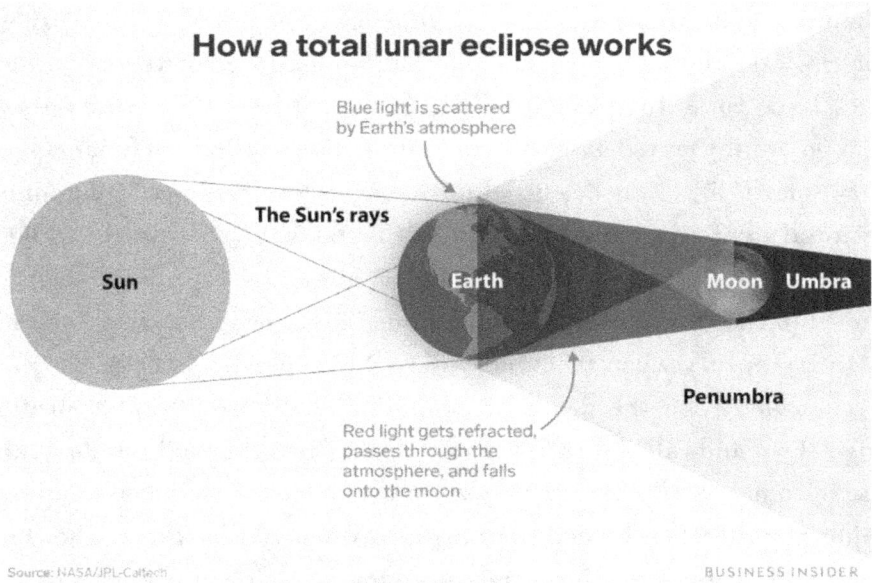

Figure 37 Lunar Eclipse- Schematic illustration.
Earth's backyard: Moon resides in the umbra to pocket the red light refracted by the Earth's atmosphere. Image sourced: Business Insider; Credits: NASA/PL-Caltech.

When the Moon is totally eclipsed, it is traversing the umbra 'the Night that shrouds' verse 84:17. It is lit by the red light clearing from the Earth's atmosphere (twilight zone) which we call as the 'Blood Moon'. After this the Moon slips to clear the penumbra—toward partial eclipse—and finally faces sunlight again. This total phenomenon for maximum duration lasts until dawn. The Arabic used for it is '**subah**' in verse 74:34. It graduates as one among many momentous portents. Quran 84: 16-18 "*I (God) swear by the twilight. By the night and what it shrouds. By the Moon as it collects. You will progress from stage to stage. So, what is the matter they do not believe?*" As and when this phenomenon takes place from night fall to dawn it is to be reckoned as traversing yet another stage toward the final Warning made by God. He has detailed this science in a beautiful way manifesting His presence and authority to dictate knowledge. Knowledge clearly is science. It is no hokum nor sorcerer's play. For this then let those who wish to step up, so do: Quran 74:38 "*Every soul will be held hostage for its actions.*"

One can imagine Makkah in 7th century. It was very impoverished patch in the desert. They did not produce men or women of insight. So, they trod ignorantly. Perhaps they cared less about night skies up until- after the Quran. Certainly, lunar eclipses traversed their skies which they hardly noticed unlike their neighbouring civilisations who also in human ignorance were apprehensive of odd celestial events. It was the Quran that just poured out of the sky. And these passages have come to speak for God. Not until now these passages weren't understood or interpreted at least in my knowledge to allude to lunar eclipses. On contrary solar eclipse was discussed in the life of the Prophet Muhammad. The longest total solar eclipse between 4000 BCE to 8000 CE a span of 12,000 years will occur on July 16, 2186 and will last 7 mins 29 sec.[227] Solar eclipse is not mentioned in the Quran like it has mentioned lunar eclipse to last from night to dawn. In 7th century when the Arabs inferred to superstition by the sun's eclipse which coincided with the death of Prophet's new born; Prophet Muhammad corrected them and called it superstition. He clarified that eclipses do not happen because of someone's death but they are among signs of God to remind us of our grand journey therefore he said to praise God when you see them. The Quran did not make solar eclipse a point of reference but moon's night long retreat in the backyard of the earth is marked as among the portents. The likelihood being as moon is the earth's time keeper. In passages of the Quran describing eclipse the technical beauty in God's scripture is shown to be outstanding. The Arabic word أفل (afala) used in verses 6:75-79 to refer to the setting of planet, moon and sun

---

[227] Source: NASA.gov.

below the horizon as discussed in section 3.1 is omitted from use in these passages referring to the eclipse. Instead appropriate technical terms are used to picture the retreat of moon in the earth's backyard.

Punishments on Arabs and Holy land being reoccupied by house of Israel are among other sings of God telling of doom. And there will come a time when this abated another warning after the previous will come to pass at which forgiveness would not be possible. As a result, faith will cease to be which anyone can earnestly seek. It is very plausible that our confirmation bias will be rock solid by then not to allow us to concede to even more clearer evidences, such as: Quran 27:81-82 *"You (O' Muhammad) cannot guide the blind to avert them from delusions: Only those will listen who believe in Our verses as they are inclined to submit. And when the Word is fulfilled against them, We (God) shall produce from the earth a creature to speak to them: that mankind were never certain of Our verses."* And any time after this event happening will eventually the Doomsayers' word come to pass. Until then the pseudo Doomsayers of our times are in fact weary of patience and perseverance for striving towards good practice. Frustrated they give up on enjoining good instead they falsify scripture to announce doom for all on a certain date which never happens. And as seen time and again their false calling of earth's imminent doom on a given date thus becomes a trail for the sceptics to earn doubts concerning the truth about God and His religion.

9.3 BREAKDOWN OF THE COSMIC SKY

> Quran 21:104 The Day when We shall roll up the Sky as a recorder rolls up a written scroll. Like We began the first creation, We (Allah) will repeat it. A promise binding upon Us. Indeed, We (Allah) will do it.

The first part of the above verse is sheer symbology. Allah rolls up the sky like the scribe rolls up a written record is an allegory. It is to mean that the recording of a grand show concerning our worldly affairs is over. Angels drop their pens and roll up records to be presented before Allah like the recorder rolling up his scrolls. Long after this unfolding, the earth will turn dead akin to its first creation. God will then undertake resurrecting life in a promise binding upon Him, "...*a promise binding upon Us. Indeed, We (Allah) will do it.*" It can fairly be fathomed in our common knowledge today of how Allah would assume reviving us from our traces of DNA.

> Quran 85:1-2 By the sky containing starry constructed mansions. And (by) the promised Day.
>
> Quran 82:1-2 When the sky **breaks open**. And when the astronomical bodies scatter.
>
> Quran 77:7-10 Indeed, what you are promised is to occur.
>
> 8. So, when the stars are dimmed (obliterated).
>
> 9. And when the sky is cleft asunder.
>
> 10. And when the mountains are blasted.

By invoking a vow upon the sky filled with starry-constructed mansions, Allah emphasises sure encounter of the promised day. This day is said to release the sky wherein the stars are intricately sewn within. The sky in 85:1 is the overarching sky of the universe encompassing skies for all individual regional dooms. And the sky in 82:1 is appropriately the first Cosmic sky in which localised doom will upset astronomical bodies. That day stars in this celestial patch will be obliterated. This ensues the expediting of our sun's star-death. Which then necessitates tearing and ripping apart of our planet earth and its skies. The sky here in 77:9 is the earth's Magnetosphere which would have turned infirm. Verses 77:7-10 communicate details of earthly sky (Magnetosphere) cleaving apart- a phenomenon to follow from when the first Cosmic sky is commanded to break open. And mountains blowing away follows after this. When the sun is set into dying. *Quran 84:1-5 "When the sky is ripped apart, Obeying its Lord as it rightly must. When the earth is levelled out. Casts out its contents and becomes empty. Obeying its Lord as it rightly must."* The jolting of the Earth collapses the core function. As a result, Magnetosphere breaks open in discrete vortexes, also see 9.5. The Earth's crust shook hard is displaced. And mountains thus are uprooted and displaced from the earth's crust. Flattening peaks, the Earth will pose a level ground that day.

> Quran 81:1-3 When the sun is wrapped up (of its light). And when the stars turn quaggy losing lustre. And when the mountains are displaced.

The Quran tells of doomsday or day of judgement as the commencing of a very long period of day. Clearly a cosmically necessitated time-period which is going to unfold in thousands of years of our count wherein the Lord of Cosmos- of the great throne, will descend making a proportional visit. With eight Angels bearing Allah's throne that day: *Quran 69:16-17 "And the sky will split, for that Day it is infirm. And the Angels are at its edges. And eight will bear the Throne of your Lord above them, that Day."*[228] It is a day which is going to last for many hundreds and thousands of years as per our reckoning of time. On that day the structure of space-time by Allah's orders will warp in our domain as the sky of the dark matter is commanded to displacement from assigned regions of celestial space. Clearly, cosmic sky of the observable universe filled with filaments of billions of galaxies is not affected in the above said doom. Inferred from many Quran verses. Hence

---

[228] Abi Dawood 4727: English trans. Book 41, Hadith 4709; reported by Jabir b. Abd Allah (Ansari) about the size of one of the Angels who bears Al-Arsh (the throne) of Allah. The distance between that Angel's ear lobe (symbolism) and his wing-shoulder is equivalent to a **seven hundred**-year journey. Viz. the journey talked about is of an Arab horse galloping at desert speeds of approximately 22 mph. In another narration, the distance to sun from earth is said to be a five hundred-year journey. The distance covered at this speed for a period of 500 lunar years is approximately (150,827,719,680 meters) very close to one astronomical unit (AU) which turns out to be the distance between Sun and Earth, as of 2012 it is (1 AU = 149,597,870,700 meters). In Arabic 70 and 700 are prominently used for numerical expression analogous to my cultural (Indian) languages where 10 and 100 are used to express exaggerations in numerical counts or to express scales. It mostly means the real measure is usually less than the expressed maxim.

cementing the view that it is a local phenomenon destined only for solar neighbourhood in our domain. Stars from this patch of sky would fall at the centre of the shifted force. This unfolding of doom for many futuristic years commences at the sounding of first trumpet. People will then get to watch the catastrophic calamity of the sky's natural fall. This phenomenon we will get to see as it will continue to conclude at the sounding of second trumpet. When we will be resurrected: *Quran 39:68 "And the trumpet will be sounded and all in the skies and earth will fall dead except whom Allah wills. Then (awhile after) it will be blown a second time, and all will stand, looking on."* As beings would be dead at the commencement of doom and then resurrected to bear judgement before Allah. Mankind by then would be collected before God. To contemplate the calamitous cosmic disarray of the unfolding doom and yet others driven to arrange in crowds, still staring in horror. Brief lights scattering, travelling from distant stars, falling off from their trajectories would still be reaching earth for observers to perceive the dreadfulness of the day. Allah will reward the foremost among us. For all our hard-sought endeavours. And He will cause our translocation to the planet deemed for afterlife. Wherefore the felicitous as well as the wretched will have their day.

> Quran 81:11-14 And when the sky is scraped. And when Hellfire
> is set ablaze. And when Paradise is brought near. Each soul will
> know what it has brought forward.

As the drama on earth at the commencement of doom would have caused death to all except whom God spares death on that day. These fortunate observers will that day see the sun on a spree of losing its splendorous light and the earth's topography transformed. And few years into it the stars falling from their orbits in a state of freefall towards the shifted centre of force. Sun and stars could be surmounting a tremendous travel to the galactic centre or the blackhole. Resurrection of humanity will wake up to too many surprises. Relationships and fan clubs having cut. As truth alone will have value: *Quran 23:101 "So when the trumpet is blown, relationships between them will fail (of no use) on that Day, nor will they ask about one another."* Resources replenishing earth will be dying and seas melting away. Then, creations, humans and jinns having been assembled, would thus mark the commencement of the phase of judgement. All matters concerning worldly affairs that were endeavoured as secrets will be made public. Then this patch of the cosmic sky will be scrapped. Hellfire lit ablaze and the gardens of Paradise brought near by the displacement of a new cosmic sky-line belonging to paradise planet. Every soul by then will have learnt what it has qualified for.[229] Following this the receivers of good records will be landed at the gates of Paradise via the highway through the cosmic sky— *Quran 51:7 "By the sky full of pathways"* and will be given salutations. Awhile the unfortunate ones would be made to land at the beds of hell and will be pushed into the scorching pit. A full recompense. As they had made a similar mockery of Allah's doctrine in their days of liberty.

These are the creation narratives of Allah—the Lord of the worlds. Providing directions to our disbelieving intellectual elite- the Atheist naturalists who are busy printing epic forms of fiction to summon perplexed masses. In the end-Universe in fact has no end. The big-crunch fair postulated is delusional. The energy and plasma needed to sustain stars and planetary systems will be infused to extend lifespans for eternity. If that is in the determination of Allah's grand plan.

---

[229] Indicative of Quranic verses 81:1-14.

9.4 DEATH OF SUN AND MOON

The judgement day is set for the appointed determinism. Which becomes necessitated due to adverse evidential human responses to God's laws of mercy on prerogatives of belief and righteous conduct. The cherisher of the world Allah will therefore conclude doom.

> Quran 54:1 The Hour has drawn near, and the moon (about to be) **struck**!
>
> 2. And if they see a sign (Quran's narrative) they turn away and say, "**blithe rhetoric**."
>
> 3. And they have denied and followed their inclinations. But for every matter is a (time of) settlement.

The Quranic revelations besides guidance are a sign of concluding doom upon the world. Mercy must be earned through diligent demonstration of faith via righteous conduct. For thirteen years of the Quran's revelation to the people in Makkah, the chiefs of tribes had only this to say that it was sorcery casted or blithe rhetoric by Muhammad. As also said by the unfortunates of today. Thus, they had denied the message of truth or in other words denied the reminder from Allah and pursued their erroneous ways. God warns and concludes- the hour has drawn near and the moon about to be struck. Which means the moon making lunar calendar for earth is about to be concluded. Closing the chapter of the grand assessment for recipients of freewill in this world.

In verse 54:1 the word '**struck**' used in past tense is often misunderstood to mean the moon has been already struck. Instead, it is a prose in the Quran to emphasise the imminent. The developing context clarifies that the sign talked of as denied in 54:3 was the revelations of Allah coming their way to remind them of their duty to Him. Which they declared as charlatan's words coming from a man whom they knew was not like this before. Therefore, they rejected Muhammad. This exact narrative is seen ratified in surah Al-Hajj, Quran 21:1-3 "*(The Hour) Has drawn near for peoples' accounts, while they are in heedlessness turning away. No* **remembrances** *come to them anew from their Lord except that they listen to it while they are at play (to it). With their hearts distracted. And those who do wrong, conceal their private conversation, (saying), 'Is this (Muhammad) except a human being like you? So, would you approach (the alleged) sorcery while you can see (that it is from a man)?'*" But again, poplar folklore among Muslim traditions of moon having split is not representative of the Quran's narrative as the Quran alone is Muhammad's miracle that God gave it to him. Moon is not the sign, instead it is the Quran which the people of Makkah had denied calling it—a plain magic. Of the miracles which they demanded from Prophet Muhammad to tell the truth of his office similar to those provided to Prophets such as Moses and Christ. Allah came back cancelling their demands reinforcing that the Quran is that miracle.

> Quran 6:35 And if their evasion is difficult for you (O Muhammad), then if you could seek a tunnel into the earth or a stairway into the sky to bring them a sign, (then do so). But if Allah had willed, He would have united them upon guidance. So never be of the ignorant.

The Quran expounds death for both sun and moon. At the stage of dying, sun would have expanded its outer layers growing up large to engulf the moon bordering earth. The sun and the planets which are at present held functioning by the sky of dark matter and energy upon its dismissal the sun will inflate quickly (in astronomical sense) whose time in our domain will be at least many years of unfolding.

> Quran 54:50-55 And Our command (to call the hour) is but (just said) once like swift as the blink of an eye.[230]
>
> 51. And (of the past) We have already destroyed your kinds, so is there any who will remember?
>
> 52. And everything they did is in written records.
>
> 53. And every tiny and big (thing) is inscribed.
>
> 54. Indeed, the righteous will be among gardens and rivers.
>
> 55. In the breadth of truth with the Sovereign God, Perfect in Ability.

At the order of God time will not delay but come to pass. The sun will begin losing its gravity potential and its shinning glory. The splendorous spacious heliosphere

---

[230] Quran 16:77 And to Allah belongs the unseen (dimensions) of the skies (of Universe) and the earth. And the command for the Hour is not but swift as the blink of an eye or even nearer. Indeed, Allah is over all things competent.

will fade away which is alluded to as 'the sun's light being wrapped up' in verse 81:1.

> Quran 81:1-3 When the sun is wrapped up (of its light). And when the stars turn quaggy losing lustre. And when the mountains are removed.

A spear of carburised ruddy darkness will surround the sun and with hardly any light in the sky the earth will be immersed in a deathly gloom.

> Quran 75:6-9 (A challenger to Prophet's message) He asks: "so When is the Day of Resurrection?"

> (Response from God) Awhile, when the vision is dazzled. And the moon darkens. And the sun and the moon are joined.

The sun would grow and expand to become more bigger and brighter such as to daze our vision in its first phase of doom. In its later phase, sun's outer layers, now less bright and stretching out darker will engulf the moon, darkening it, and hardly any shine be left for it to reflect. This marks the collapse of lunar calendar and therefore the end of the planetary time. Where the scientists are unsure the Quran informs them that there will be merging of the sun and moon.

In the sky of this world, the stars in solar neighbourhood, along with the sun will be dimmed losing their shine, obliterated, and, felled indicative of verses: *Quran 77:8 "So when an-nujum (the stars) are blurred (or obliterated)"* plus *Quran 81:2 "And when an-nujum (the stars) turn quaggy losing lustre."* And lastly *Quran 82:2 "And when the kawakibu (stars plus non-luminous bodies) fall, scattering"*, they experience an erratic fall. Nothing of this is suggestive of stopping celestial cruises of astronomical objects albeit it is only their regular orbital cruises that are lost. Therefore, they do not attain resting points in spatial grids in the Cosmos, but these bodies will continue to fall scattered towards the centre of the shifted force in the direction of star setting sites or the centre of the galaxy.

Earlier mentioned verse references (see chapter 2), such as 35:13, 31:29 and 39:5 and *Quran 13:2 "...each **running** (its course) for a specified term..."* talk about the heavenly bodies in orbital cruises. The Arabic verb تجري (tajri) is traditionally translated as **running** courses. Technically it is meant for **orbital cruises**. In verse 36:38 تجري (tajri) precedes a passive participle لمستقر لها (li-mustaqhar laha) meaning -cruise as determined for it (sun)- similar to verse 13:2 لأجل مسمى (li-

300

ajalim-mussamma) meaning -cruise until appointed or specified term-. [231] In modern Quran translation of Sahih International; the mistranslated part of verse in 36:38 as a -stopping point- for لمستقر لها (li-mustaqhar laha) is implied by the critics as the only correct translation of this phrase. For inducing the meaning of perhaps coming to a complete physical stop of the astronomical bodies. In fact, this bit is an indicator for a specified physical determinism of its orbital cruise. The Quran does not suggest stopping at all. In verse 36:38 لمستقر لها (li-mustaqhar laha) meaning -as determined for it (sun)- cannot mean stopping as in a physical stop. The passive participle in verse 36:38 is a determination impending for the sun's and the stars' catastrophic doom phenomenon as described in previous paragraph for their dimming and scattering. It is a determinism or an estimate of term which indicates compatibility with its preceding verb i.e. it's orbital cruise. As the verb and its passive participle do not contradict as in to cruise and stop therefore the syntax is explanatory of bodies in float through the celestial space and never stop as misunderstood by eager enthusiasts. Instead as highlighted earlier regular orbital cruises perhaps will fall apart leaving them in scattered orbits to the quasar in the galaxy centre for full annihilation without any intermediary physical stoppage.

> Quran 6:33 "We (Allah) clearly are aware the grief their words
> do cause you: It is not you they reject: it is the verses of Allah
> which the wicked contend."

Gone are the days when the ancients had come to think that the Sun-god was all powerful. Today sun is dwarfed for its little power in comparison with distant stars. And clearly the thought of it as a god stands abandoned. Why must the Quran above all come to decipher the workings of astronomical bodies in bold details and yet be a forgotten relic to modern masses who even today praise heretic ancients for science inquiry which has produced many half-baked postulates. While Muhammad's unchallenged contributions as shown in this work on astronomy are necessarily alleged as wrong and stand forgotten. Clearly as said by God Himself: *'it is not you O' Muhammad they reject but it is God they-*

---

[231] Quran 36:38 And the sun runs (its course) as determined for it (sun). That is the decree of the Exalted in Might, the Knowing.

disbelievers contend with'. The fact is if Prophet Muhammad were not a Prophet he would be unknown in relation to the Quran- the troublesome words of God.

9.5 WEAKENING OF EARTHLY SKIES

> Quran 35:41 Indeed, Allah holds the skies and the earth, lest they cease. And if they should cease, no one could hold them (in place) after Him. Indeed, He is Forbearing and Forgiving.

Allah has demonstrated His signatures manifestly within our bodies and in the near horizons of the Cosmos. Tellingly that He and none other than He bears the attributes of origins, functional creation, and, artistry in His domain. He alludes to earth and its skies as being held and meticulously administered by Him for its proposed operations. By inferring to examples of what could result lest He abandoned them He shows us in our near habitation within the solar system the parables of His waiver. New learnings from planet Mars's historical past come close to explaining this parable. Mars slightly smaller than earth still shows signatures of ice caps at its poles and satellite imagery shows evidence of rivers once channelled on its surface. It became desolate as it couldn't contain its atmosphere supposedly due to Iron core failure causing its water to evaporate turning it into a red desert. Then mesmerising Venus is rather lethal with the planet engulfed by peak greenhouse effect; dominating carbon dioxide gas increases its temperatures to staggering levels and thick gas clouds rains acid. Both Venus and Mars were most likely to hold life because of their readiness and residence within goldilocks zones of our sun: *Quran 35:41 "Indeed, Allah holds the skies and the earth, lest they cease."* Allah perfects from His selection of what gets resulted numerously in the making by His command. If Allah were not to uphold our planet to duty, then earth and its skies would cease to function. Then there is none after Allah to restore them back to a working unit. Allah alone forbears and forgives much from many sinners among us despite our rebellion to Him.

> Quran 17:44 The seven skies and the earth and whatever is in them exalt Him. And there is not a thing except that it exalts (Allah) by His praise, but you do not understand their (way of) exalting. Indeed, He is ever Forbearing and Forgiving.

Allah drives home the point that everything in His creation nearly surrounding us exalt Him. Except many numerous recipients of freewill. The seven skies and the earth and all that is in them exalt Allah. But we don't comprehend their exalting. Indeed, Allah forbears and forgives much of our rebellion to Him. And yet He by His mercy guides us to light even as miserly of will we stand to reform.

Quran 69:13-15 Then when the Horn is blown with one blast. And the earth and mountains are jolted and levelled with one blow- Then on that Day, the encounter (Resurrection) will occur.

When only rebellion to Allah prevails as the dominant religion (way of life); then God will at the onset of the appointed determinism for judgement day, the cherisher of the world Allah, will therefore conclude doom. Awhile when the horn is blown by a near earth bystander Angel *Quran 50:41 "And listen on the Day when the Caller (Angel) will call out from a **place nearby**"*, the earth is jolted to violent shaking and gripped into convulsions crumbling the mountains under strong pressure waves and blowing them away.[232] Detail of verse 50:41 i.e. from a **place nearby** where the call can be heard to earthly beings is sizzling. There was no need for the verse to conclude with the Arabic phrase من مكان قريب (min makanin qharib) translated 'a place nearby'. This is the beauty in Allah's narrative. He draws attention to subtle facts in this way—simply elegantly. Telling the science of facts. In this case the need for an atmospheric medium for propagation of sound waves. Hence the Angel for this job hangs out lurking very close to earth such that his blow of trumpet will blast through the medium of air globally.

Quran 79:1-14 By the forceful (Angels) who tear out (wicked souls). By those (Angels) who gently draw (souls of blessed). And those who glide in spaces. Then who press forward in race. Then bring the matter to an end. One that day (earth will quake) in violent commotion. Followed by oft-repeated (commotions). Hearts will tremble. Eyes will be downcast.

These (presently) say, 'What! shall we be brought back to life? After we have turned into decayed bones?' And these concede: such a return will amount to loss! But verily, it will take only a single blast. When they will be back fully awake.

That day resurrection will occur. Due to earth's convulsions, its Magnetospheric sky will be flimsy as the earth's core would have become partly dysfunctional due to jolting of the earth from an elusive dimension. This calamity will make other

---

[232] Quran 99:1-3 When the earth is shaken to her (utmost) convulsion. And the earth throws up her burdens (from within). And man cries (distressed): 'What is the matter with her?'

skies encompassed within Magnetospheric sky tear apart in vortexes of Magnetosphere splitting and displaying a rose like hide: *Quran 55:37 "And when the sky is split, and it becomes rose like hide."* The context in the above set of verses is indicative of the Magnetospheric sky becoming infirm resulting in the tearing apart of all encompassed skies including near atmosphere (troposphere) with clouds de-clustering. Angels would assume their positions at its edges: *Quran 69:16-18 "And the sky will split, for that Day it is **infirm**. And the Angels are at its edges. And eight will bear the Throne of your Lord above them, that Day. That Day, you will be exhibited (for judgement); not hidden among you will be anything concealed."* Again, the Quran does not indicate complete annihilation of earthly skies but tells that they will be made **infirm**. Which is a most likely happening, evident from observations of examples set in our neighbouring planetary bodies such as Mars and Venus suggest similar tails.

The atmosphere of Homogenous and Heterogenous skies will sheen rose like hide. Rose colour spectrum is implied to blossom that day as the Quran uses the phrase وردة كالدهان (wardatann kad-dihaan) meaning like rosy splash of lipids or rosy smear of paint translated as rose like hide in the above translation. A further metallic tinge of molten copper or that of brass is implied to smear the expanding and tearing sky to be seen in the visible spectrum: *Quran 70:8-9 "On the Day the sky will be like murky oil. And the mountains will be like wool."* In describing the status of the sky surrounding earth Homogenous and Heterogenous layers will partake in respective spectral petals to resemble bloom similar to a rose hide. Awhile mountains will be powdered and thrashed like carded wool.

> Quran 52:5-10 And (by) the sky raised high. And by the Ocean filled with swell. Verily, the Doom of thy Lord will indeed come to pass. There is none to avert it.
>
> On the Day, the sky will sway with circular motion. And the mountains will pass on, departing.

Heterospheric sky (encompassing Plasmasphere plus other imbedded layers) alluded to in the above verses by its contextual association to sweltering oceans is said to sway in circular motions. It is indicated that due to collapse of Magnetospheric sky at it becoming flimsy because of abnormal operations of the Iron core; as the collapse of magnetic dipole. It breaks and tears the blanket surrounding earth into broken vortexes. This then bears upon the Plasmaspheric sky co-rotating with it. Which rolls the matter in it in whirls along the magnetic field lines.

> Quran 25:25 And the Day the sky shall be rent asunder with the clouds, and angels shall be sent down, descending (in ranks).
>
> 26. That Day, the dominion as of right and truth, shall be (wholly) for (Allah) Most Merciful: it will be a Day of dire difficulty upon Misbelievers.
>
> 27. The Day that the rebel will bite at his hands, he will say, "Oh! Would that I had taken a (straight) path with the Messenger!

Homogenous sky (Homosphere plus imbedded layers) will tear open due to the opening of the enclosure i.e. Magnetospheric sky. Due to this effect, the Homogenous sky will experience a rapid expansion. It will disperse its tightly held gases along with its clouds tearing apart displaying colours. About this time, Angels will descend in arrays upon arrays to grip beings to summon them before Allah. That day to Allah will belong the Kingdom as the recipients of freewill will be forced to surrender before Allah, the Glorious. That day, upon the rebels, it will be a tough day indeed.

> Quran 14:48 On the Day the earth will become a different model earth, and the skies (as well), and all creatures will come out before Allah, the One, the Irresistible.
>
> 49-50. And you will see the criminals that Day bound together in shackles. Their garments of liquid pitch and their faces covered by the Fire.
>
> 51. So that Allah will recompense every soul for what it earned. Indeed, Allah is swift in account.
>
> 52. This (Quran) is a notification for people, such that you be warned thereby and that you may know that He is but one Allah. And such as those of understanding be perhaps reminded.

In a crux all earthly skies would gradually be withered apart. Once honoured to cherish life they will then become weak-modelled skies. And the earth itself will pose like a flattened surface, nil of slopes, troughs or valleys. With mountains crumbled and gone, it will that day be a different earth, with only open plains to see. Criminals before Allah will be busted and bound in shackles. Garments upon them will be of tar and their faces charred by the fiery smoke. Such must Allah recompense every soul for what it efforted to work for, for Allah is as simply said, just and fair.

No matter how far we progress to opinion Allah's non-existence after having failed to learn from His accounts given to earlier civilisations, we have chosen to remain clueless of what is to befall us:

> Quran 30:25 "And of His signs is that the sky (Homosphere) and earth stand by His command. Then when He calls you with a (single) call from the earth, immediately you will come forth."[233]

For God to pronounce judgements on that day which is a long day (perhaps thousands of years); His creation thus resurrected needs essentials of air. Then as rightly alluded, the Homosphere will stand by His command for expediting finalities for believers and disbelievers alike. They will be called from their earthly depositories back to life to receive their reports and appraisals with God.

The last of Allah's accounts delivered by Angel Gabriel to Prophet Muhammad is found even today in its original recital. It is meant to be in its original till this earthly trial concludes. A revelation containing remembrances of the Lord of the worlds. That which Allah calls it—a clear proof! With it, it is like you're learning in Allah's own presence. The biosphere where all life exists is called in verse 30:25 which is the Homospheric (sky) plus the earth's lithosphere. Allah informs that this unit will stand by His command after the first blow of trumpet and at length when He will announce the call calling us at the second blow of trumpet: Lo! we will have resurrected back to life thus we will respond to the call and others be driven by stern angels to march in arrays before God that day. Isn't this a detail that supersedes our science perceptions of what and how the planet will be systemically broken-down?

---

[233] Context of the sky a singular noun talked in verse here is Homosphere. Also see, Quran 30:24 And of His signs is (that) He shows you the lightening (causing) fear and aspiration, and He sends down rain from the sky by which He brings to life the earth after its lifelessness. Indeed, in that are signs for a people who use reason.

Unique legions of Angels varied in sizes would arrange themselves at the boundaries of Magnetosheath. Still above most credited Angels eight of them will bear the throne of Allah as humanity and jinns be marshalled to present before Him. Scores of Angels will further descend upon the earth in ranks to drag criminals before the Lord of the worlds. Then it will be told: *Quran 37:21 "This is the Day of Sorting out, whose truth ye (once) denied!"* From the looks of it (from **Figure 36**) it is going to be a close theatre!

Either God exists and reveals truth. Or there is no God. Because God is not mute and cannot be the proponent of falsehood. If there is no afterlife and therefore no punishments or rewards, then karma is falling off its foundations. And this means we could do whatever we could in pursuits of desires without worrying for any futuristic consequences and change laws by power and be free. This therefore invalidates God's ability to support truth over falsehood. Which means God is irrelevant and thus not-required. If all this is false, then God is the relevant truth and He will judge us for everything. Even if this means that He will translocate us to a new planet untold of its exact celestial location to meet His promises made of punishments and rewards. Hence this means human desires to bypass God's commands will not go unpunished. There is a clear degree of match in all fundamental concepts, faith in One God, belief in afterlife, belief in Angels, Prophets and Books of God found in Vedic scriptures, the Bible and the Quran. [234] Teachings with similar foundations but differing apart due to history, place, time and language.

It is not the peer-approval or struts of being elite, or the outfit of a scholar deriding humaneness or the claim of knowledge that will pay as despite these one may pursue a material gain. It is but, (Arabic: Ikhlass) sincerely attained deeds to please Allah Almighty and a just bias at one's bare minimum that will truly pay. Knowledge and righteous actions are always hand in hand. Prophet Muhammad is famously quoted: *even a smile could be charity!* As in Quran 2:82 *"But those who believe and do righteous deeds- they are the companions of Paradise; they will abide therein eternally."*

The benefit of fighting scepticism for faith is to earn our Lord's good pleasure and to accrue rewards as great as lasting youthfulness. In a place teaming with never exhausting enjoyments for which Allah has setup up three simple check points: If any successfully demonstrates sincerity to Allah, testifies that no other claim

---

[234] Equivalent derivatives from Vedic scriptures for Angel is *devdhoot*, end-times is *kali yuga*, Satan is *asura* also nick named *pishash*, doomsday is *pralaya*, eternal life is *moksha* and Hell is *naraka*.

of god is worthy of any trust and worship but God, the Creator and then commits to good work, such an endeavour stands to be acknowledged by Allah.

> Quran 2:62 Those who believe (in the Quran), and those who follow the Jewish (scriptures), the Christians and the Sabians (truth seekers); any (among mankind) who believe in Allah, the Judgement before Him and works righteousness (doing good) shall have their reward with their Lord. On them shall be no fear, nor shall they grieve.

Regardless of subscription to any religious idea, belief in Allah alone as- the God, His Judgement, and doing good actions will never let us to suffer in afterlife.

> Quran 3:110 You are the best nation produced (as an example) for mankind. You enjoin what is right and forbid what is wrong and believe in Allah. If only the People of former Scripture (Jews and Christians) had believed (now in Quran), it would have been better for them. Among them (yet) are believers, but most of them are defiantly disobedient.

Fourteen centuries past today, this verse here read afresh condemns many cultural Muslims of today for their characteristics of being defiantly disobedient like the many dissenting factions of Jews and Christians after guidance had reached—who had become blinded. This explains the many ills wrought by religious people to paint God with a brush full of ignorance and sin. We are in times of a prophecy told by Prophet Muhammad: *There will come a time upon mankind, when nothing will remain of the Quran except its custom. Their mosques will be full, but destitute of guidance; their learned men will be the worst people under the sky; and contention and strife will issue from them, and it will return upon themselves.*[235] Then imagine who and how many today have the keys to heaven?

---

[235] Baihaqi; Mishkat-Ul-Masabeeh translated by Capt. A.N. Matthews, Book II, page 96. Reported by Ali ibn Abi Taulib.

Being born in the framework of justices; and while we have deduced no continuity of purpose via our extensive scientific inquires, except that we must make modest meanings of our present lives and then be doomed to oblivion. In the wake of this, can we then call in to question the credibility assigned to teleology in theology?

In answering this question: the Quran gives one straight answer by reminding us of our forgotten purpose-

> Quran 51:55-56 And remind (O Messenger), for indeed, the reminder benefits the believers. And I did not create the jinn and mankind except to worship Me.

Worship told in verse 51:56 here is not just praying at certain times of the day. But, an overall conductance of righteous life to impress Allah.

For those people who aren't worried to consider Allah an option, won't the unfolding for them become precarious? As they are to most likely face His vengeance. Having lived a life which is by God's special grant, those who have taken to opinion His non-existence or it as a thing simply redundant to consume any worthy time of consciousness, it will be a huge confrontation at their passing away, when they will get to experience the unseen world of Angels who are surely entrusted to expedite procedures of dying. In fact, it is like to piece a cake, to recognize Allah as our creator. And for some it is worthy a contemplation to read a critique of the opposing views to see God as clear as day. If any has rejected the proposition of God's non-existence citing premises of credible evidence having not surfaced by our scientific enquiry, he has thus favoured belief in God's existence. This principle in my opinion is simplest of all to make situation look far from precarious before God.

> Quran 39:53-54 Say, "O My servants who have transgressed upon themselves (by sins and disbelief), do not despair of the mercy of Allah. Indeed, Allah forgives all sins. Indeed, He is the Forgiving, the Merciful."
>
> And return (in repentance) to your Lord and submit to Him before the punishment comes upon you; then you will not be helped.

It is plain to know why the concept of Allah haunts us so much. Because it is a connect made deep within. As young neurons assemble in the wombs Allah takes an oath from us in which we conceded to faith.[236] This is precisely the reason- even if human upbringing is done in seclusion to the found cultures having to discuss the know of God, yet reflections concerning God are inevitable. The questioner would explore towards our beginnings and no amount of supplement theories are potent to put this enquiry to rest. Richard Dawkins an acclaimed atheist and an outspoken critique of God and religion for this reason precisely shows disability to break away from this neuronal connect. It is evident from his own reflections as he talks concerning an encounter with God that if it turns out to be for him after he dies, he will then submit. Surrounded in confusion he greets the Creator by names other than Allah's, from Zeus to Jesus. In other debates, he considers being reminded of God by the sheer take of creationary display.[237] Yet Richard Dawkins presses his famous assertive trivial though that 21st century science has achieved enormous understandings even of the near origins of life and of Cosmos, remarking however there are gaps that need filling. Deluded he comes to think of science as the major emancipator worthy of manifestation enough to disqualify belief in God. The creed from Allah like science has anciently clarified to people to turn away from earthly objects of worship. It was never expected of modern societies to relinquish faith in Almighty citing science as evidence, at least per the Quran. According to the Quran, many previous nations about what they had achieved of intellect and works, whom Allah had destroyed, had also similarly presented rebellion when called to faith. Science and research may not qualify as worthy premise to disbelieve but ideas of liberation and freedom has trickled many to enjoy godlessness.

---

[236] Quran 7:172-173 As and when your Lord summoned the children of Adam reproducing from their loins and made them testify of themselves, (saying to them), "Am I not your Lord?" They said, "Yes, we have testified." (This) lest they should say on the day of Judgement, "Indeed, we were of this unaware." Or (lest) they must say, "It was only personalities invoked by our fathers (besides Allah) and we were but descendants after them. Then would You destroy us for what the whimsical have done (to us)?"

[237] Richard Dawkins vs John Lennox | The God Delusion Debate; YouTube. Watch 21-25 minutes.

Cognitive dissonance is typical of humans and jinns. But general cognition due to consciousness is observable in almost all living species. However, this knowledge of how many living organisms including microbes organise cognition needs research. Particle physics has opened new facets for research- it is observed even particles demonstrate elusiveness such as observed in a double slit experiment.[238] In terms of acknowledging Allah's sovereignty, the Quran considers all matter alive or otherwise observes worship to the Lord of cosmos. The hymn is repeated on several occasions in the Quran such as: *Quran 62:1 "Whatever is in the heavens and whatever is on the earth is exalting Allah, the Sovereign, the Pure, the Exalted in Might, the Wise."*

> Quran 22:18 Do you not see, to Allah prostrates whatever is in the heavens and whatever is on the earth and the sun, the moon, the stars, the mountains, the trees, the moving creatures and many of the people! But upon many the punishment has been justified. And he whom Allah humiliates – for him there is no one to bestow honour. Indeed, Allah does what He wills.

Again, dissonance is predominantly human who are the responsible players on earth. Aren't we given to justifying? Clearly, a phenomenon not observed in non-human species such as in flora and fauna. Humans are prone to self-tasking the responsibility to erect justice within societies, for which they constantly pursue reasoning. Say, for example if the need to justify was put aside then the barrier separating truth and falsehood would crumble. In such a case life would become very reflexive with only rudimentary laws governing human behaviour. This is true and observable in human societies where some are compared to animals as they lack the potential to build character upon enquiry. The case wherein an argument for legal advocacy of homosexuality is put forward after observing homosexual behaviour in animals is childlike. Lawrence Krauss above all could have brought a better debating point to stage in his contributing capacity for this incredible issue. If asked whether on this premise if advocacy of polygamy and paedophilia by consent or the typical of jungle rogue behaviour or even seducing neighbourhood wives to broaden one's prospects of siring is justified, I'm certain that Lawrence Krauss would exclaim preposterous! Then, shouldn't the first

---

[238] That which explains the dual nature of photons, atoms and even molecules under limits of its size.

analogous premise be also retracted? If sensible enough one would concede dissonance by rationale. But reasoning is relative. Preferences interwoven in traditions and evolving cultures drive our societies. Truth-finding itself is a perpetual activity. To conclude Allah's non-existence with no burden to shoulder proof is sheer arrogance. At best one can be a sceptic or an agnostic but never an atheist. Cognitive dissonances are resolved upon good evidences like facts or rationale. It can be logical or empirical. But good evidence can always be better.

In a case in Nairobi National Park with Maasai Warriors in Kenya; faced with the traditions and strong traits of Lion hunting for several reasons other than just to defend their livestock, they were driven into cognitive dissonance. Among several measures considered to induce dissonance due to the declining Lion population were compensations for loss of livestock and education to conserve nature. But, then eventually law to deter by force has had outstanding success within Maasai Warriors subject to traditions. Anything life threatening including incarcerations have been a functional means to induce deterrence in human psychology.[239]

When a society exhausts conscionable means to educate and dissonance has not yet occurred then it resorts to introducing force to induce dissonance. This is fair. As it well seems to be a natural outcome to administer goodness. God similarly has threatened men against disbelief by setting up a place of Hell to induce dissonance. If Allah did not disclose in scripture such facts which make means to deter men from disbelief, then men would label Allah as unfair anyway for not disclosing information prior to their judgement.

> Quran 7:179 Indeed He has let sprung for Hell many of the jinn and mankind. They have hearts, but they dare not to understand, they have eyes, but they do not see, and they have ears, but they do not hear. Those are like livestock; rather, they are more astray. It is they who are the heedless.

The teleology of theology is that Allah is the final judge and all matters and disputes return to Him and in the end, justice will be served. No one will be

---

[239] National Geographic Documentary - Lions vs Maasai Warriors - Wildlife Animal; YouTube.

wronged by the Knowledgeable and the Wise God. Can anyone deduce a conclusion on Allah's non-existence in absence of clear knowledge?

> Quran 50:28-29 (Allah) will say, "Do not dispute before Me while I had already presented to you the warning. The word will not be changed with Me and never will I be unjust to the servants."

> 30. On the Day We (Allah) will say to Hell, "Have you been filled?" and it will say, "Are there some more."

> 31-34. And Paradise will be distanced near to the righteous, not far. (It will be said), "This is what you were promised – for every returner (to Allah) and keeper (of His covenant), Who feared the Most Merciful unseen and came with a heart returning (in repentance). Enter it in peace. This is the Day of Eternity."

> 35. They will have whatever they wish therein, and with Us is more.

If God did not institute test for the recipients of freewill then evidential findings of dissonance would be a contradiction. In other words, if teachings from God did not allude to this teleology in that case cognitive dissonance would infer a contradiction of faith propounded by Allah the Lord of the worlds. This teleology aimed at recognising Allah and then to be fair and just is instrumental to successes in afterlife.

> Quran 11:123 And to Allah belong the unseen (aspects) of the heavens and the earth and to Him will be returned all matters, all of it, so worship Him and rely upon Him. And your Lord is not unaware of that which you do.

How can one justify the claim made by theology concerning teleology as true? In other words, how proximate to rightness can justified true beliefs be considered. The reality of philosophy other than agreed upon by consensus is perhaps subject to relative perception and is always subjective, therefore knowledge thereof is perpetual. Some true beliefs like some falsehoods doesn't need justification as they are overly clear to common consensus although practice of falsehood is differed to discretion. But it is wise to time and again ratify beliefs for truth worthiness via pursuing knowledge. Justification is only attainable up to what knowledge has been perceived. Justification to practise a set of clauses then must at best be clear from association of lows: the range of lows can be practice of falsehood or bad-logic or preference of weaknesses over truth and nonetheless preference of kin, kith and race over justice etc. If then one assumes a belief which directs him towards goodness, then what ensues is necessarily true as it evades falsification of it. However, certainty of true cognition can only occur at the expense of the probability of dissonance. In informed theology, this probability is the test which we endure as part of Allah's plan to sort the best among us. Truth and certainty will then prevail in what is identified as afterlife post resurrection. In surah Al-Haqq (Chapter 69: The clear truth) Allah—The absolute truth, informs us about the certainty yet to unfold for our complete dissonance of falsehood.

The purpose of life is that we acknowledge God and hope to fortify our beliefs by believing that we will meet Him one day. This can be achieved by a solitary endeavour or by organising our societies for bringing forth collective goodness for this exact purpose.

About the Author

Masood Ahmed,

The author is a student of religions. In his mid-thirties, he is focused to tackle contemporary challenges in theology. He was bred of an Abrahamic belief system and has a thought-provoking grasp on Biblical and Islamic narratives. Well versed in undertaking a clear analysis of the language of the Quran. He has a bachelor's degree in Mechanical Engineering and has enjoyed a long season of practice in designing commercial aircraft structures. He is now discovering that writing to advance the cause of God is his greatest passion.

VISIT

https://allahsnarrative.com/
Instagram: @masood.aussie
"Creation of Earth and Origins of life-Allah's narrative" is next.

For permissions to use material from this product, publishing rights, franchise for sales of this book, for convertibles into other media and for translations please email- authorofancientastronomy@gmail.com with subject line- Permissions. And for criticisms, praise or any other feedback: Please email with subject line- Feedback.

Index

Z

Zanjabil · 94
Zeal · 18, 235, 282
Zero

Cusps · 17, 258
Protoplanets formation · 156
Thickness correction · 118
Time of nothingness · 82, 142
Two bows spacing · 236
Zeus · 312

www.ingramcontent.com/pod-product-compliance
Lightning Source LLC
Chambersburg PA
CBHW081554220526
45468CB00010B/2654